Workshop on Condition Assessment Inspection Devices for Water Transmission Mains

About the Awwa Research Foundation

The Awwa Research Foundation (AwwaRF) is a member-supported, international, nonprofit organization that sponsors research to enable water utilities, public health agencies, and other professionals to provide safe and affordable drinking water to consumers.

The Foundation's mission is to advance the science of water to improve the quality of life. To achieve this mission, the Foundation sponsors studies on all aspects of drinking water, including supply and resources, treatment, monitoring and analysis, distribution, management, and health effects. Funding for research is provided primarily by subscription payments from approximately 1,000 utilities, consulting firms, and manufacturers in North America and abroad. Additional funding comes from collaborative partnerships with other national and international organizations, allowing for resources to be leveraged, expertise to be shared, and broad-based knowledge to be developed and disseminated. Government funding serves as a third source of research dollars.

From its headquarters in Denver, Colorado, the Foundation's staff directs and supports the efforts of more than 800 volunteers who serve on the board of trustees and various committees. These volunteers represent many facets of the water industry, and contribute their expertise to select and monitor research studies that benefit the entire drinking water community.

The results of research are disseminated through a number of channels, including reports, the Web site, conferences, and periodicals.

For subscribers, the Foundation serves as a cooperative program in which water suppliers unite to pool their resources. By applying Foundation research findings, these water suppliers can save substantial costs and stay on the leading edge of drinking water science and technology. Since its inception, AwwaRF has supplied the water community with more than $300 million in applied research.

More information about the Foundation and how to become a subscriber is available on the Web at **www.awwarf.org**.

Workshop on Condition Assessment Inspection Devices for Water Transmission Mains

Prepared by:
Kevin Lillie, **Christopher Reed**, and **Mark A.R. Rodgers**
Northumbrian Water Technical Centre Limited
(formerly Northumbrian Lyonnaise Technology & Research Centre Limited)
Horsley, Newcastle upon Tyne, NE15 0PA, United Kingdom
and

Simon Daniels and **David Smart**
Issquared Limited
Riverside Suite, Business & Innovation Centre, Sunderland Enterprise Park,
Wearfield, Sunderland, SR5 2TA, United Kingdom

Sponsored by:
Awwa Research Foundation
6666 West Quincy Avenue, Denver, CO 80235-3098

Published by:

Awwa Research Foundation
Advancing the Science of Water®

IWA Publishing

DISCLAIMER

Published by IWA Publishing, Alliance House, 12 Caxton Street, London SW1H 0QS, UK

Telephone: +44 (0) 20 7654 5500; Fax: +44 (0) 20 7654 5555; Email: publications@iwap.co.uk;
Web: **www.iwapublishing.com**

AwwaRF report number 91028F. Originally published by AwwaRF for its subscribers in 2004.
IWA Publishing version published 2005.

ISBN 978-1-84339-916-2

CONTENTS

TABLES ix

FIGURES xi

FOREWORD xv

ACKNOWLEDGMENTS xvii

EXECUTIVE SUMMARY xix

CHAPTER 1: INTRODUCTION 1
Background 1
Objectives 1
Project Approach 2
Consultation 2
Workshop 3
Web site 3

CHAPTER 2: BUSINESS NEEDS REVIEW 5
Introduction 5
Business Drivers 5
Condition Assessment in Other Industries 5
High Level Drivers within the Water Industry 10
Condition Assessment and Asset Management 15
Business Justification 20
Deferral of Capital Expenditure 20
Reduction of Capital Expenditure 21
Reduction of Operating Expenditure 22
Economics of Condition Assessment 25
Business Requirements 27
Introduction 27
General Market 27
Failure Modes of Pipelines 28
Utility Strategies 28
Telephone Consultation 30
The Market for Condition Assessment 30

CHAPTER 3: TECHNOLOGICAL REVIEW 39
Introduction 39
Previous AwwaRF Projects 39
Nondestructive Testing of Water Mains for Physical Integrity [Project #507] 39

Electromagnetic Inspection of Pre-stressed Concrete Pressure Pipe [Project #2564] ... 40
Nondestructive, Noninvasive Assessment of Underground Pipelines [Project #355] ... 41
Techniques for Monitoring Structural Behavior of Piping Systems [Project #2612] ... 41
Testing and Condition Assessment of Joints in Water Distribution Pipelines [Project #2689] ... 42
Nondestructive Testing Techniques ... 42
Traditional Techniques ... 42
Utility A ... 44
Utility B ... 44
Utility C ... 45
Conventional Technologies ... 45
Identifying and Selecting Technologies ... 62
Manufacturers and Suppliers ... 64

CHAPTER 4: NON-INTERRUPTIVE CONDITION ASSESSMENT REQUIREMENTS .. 67
Introduction ... 64
Target application ... 67
General Condition Assessment Requirements ... 67
Outputs Required ... 68
Spatial Requirements ... 69
Operational Conditions ... 69
Impact on Levels of Service ... 69
Network Control ... 70
Field Operations ... 70
In-Pipe Systems ... 71
Assessment Parameters ... 71
Requirement Specification ... 74
Needs Not Addressed ... 74
Outline Specification ... 78
Device Operational Features ... 78
Pipe Parameter Measurement ... 78
Degradation Parameter Assessment ... 79
Other Parameters ... 79

CHAPTER 5: CONCLUSIONS AND RECOMMENDATIONS ... 81
Conclusions ... 81
Recommendations ... 82

APPENDIX A: COMPANIES CONSULTED ... 85

APPENDIX B: OTHER COMPANIES INVITED TO PARTICPATE IN THE CONSULTATION ... 87

APPENDIX C: UTILITIES INVITED TO PARTICIPATE IN THE WORKSHOP 89
APPENDIX D: LIST OF DELEGATES 91
APPENDIX E: WATER UTILITIES' FEEDBACK 93
APPENDIX F: TECHNOLOGY PROVIDERS' FEEDBACK 95
APPENDIX G: CONSULTANTS & ACADEMIA FEEDBACK 97
APPENDIX H: CASE STUDY – WATER UTILITY 99
APPENDIX I: CASE STUDIES 103
APPENDIX J: MANUFACTURERS AND SUPPLIERS 129
REFERENCES 131
ABBREVIATIONS 137

TABLES

2.1 High and low pressure pipelines 8

3.1 Acoustic emission monitoring 46

3.2 Advantages and limitations of acoustic emission technique 46

3.3 Acoustic leak detection 47

3.4 Advantages and limitations of acoustic leak detection 47

3.5 Broadband electromagnetic 48

3.6 Advantages and limitations of broadband electromagnetic technique 49

3.7 Optical – closed circuit television 49

3.8 Advantage and limitation of optical technique 50

3.9 Long range ultrasonic 50

3.10 Advantages and limitations of long range ultrasonic technique 51

3.11 Magnetic flux leakage 51

3.12 Advantages and limitations of magnetic flux technique 52

3.13 Remote field eddy current 53

3.14 Advantages and limitations of RFEC technique 54

3.15 Remote field eddy current – transformer coupling 55

3.16 Advantages and limitations of RFEC/TC technique 56

3.17 Seismic pulse echo 56

3.18 Advantages and limitations of seismic pulse echo technique 57

3.19 Ultrasonic 57

3.20 Advantage and limitation of ultrasonic technique 58

3.21 Water pipeline feature detection and measurement capabilities 58

3.22	Summary of technologies versus assessment parameters	62
4.1	Potential pipe failure mechanisms	68
4.2	Parameter values	71
4.3	Pipe parameters	72
4.4	Degradation parameters	72
4.5	Other pipe aspects parameters	73
4.6	Operational parameters	73
4.7	Summary of assessment parameters versus technologies (metallic pipe)	75
4.8	Summary of assessment parameters versus technologies (mineral based pipe)	76
4.9	Summary of assessment parameters versus technologies (plastic pipe)	77
D.1	List of delegates who attended the workshop	91
E.1	Logistics	93
E.2	Structure and content of programme	93
E.3	Conclusions	93
F.1	Logistics	95
F.2	Structure and content of programme	95
F.3	Conclusions	95
G.1	Logistics	97
G.2	Structure and content of programme	97
G.3	Conclusions	97
I.1	Land use category assignment (LUC)	109

FIGURES

2.1 Risk model ... 9
2.2 Lost water cost for various operating pressures ... 22
2.3 Break even chart ... 23
2.4 Net Present Value comparison ... 26
2.5 Composition of pipe networks ... 31
2.6 Major problem areas of pipe networks ... 31
2.7 Smallest transmission main diameters ... 32
2.8 Largest pipe diameters ... 32
2.9 Levels of understanding of the condition of pipe networks ... 33
2.10 Current practices of condition assessment ... 33
2.11 Interest in condition assessment ... 34
2.12 Appeal of condition assessment in the water industry ... 34
2.13 Awareness of NDT in the condition assessment of water pipes ... 35
2.14 Activity of condition assessment companies in the water industry ... 35
2.15 Activity of the water industry approaching the condition assessment industry ... 36
2.16 Type of tool required ... 37
2.17 Expenditure on condition assessment ... 37
3.1 Schematic showing insertion of SoundPrint® hydrophone array into water main ... 46
3.2 Schematic showing configuration of hardware for acoustic leak detection ... 47
3.3 Picture of Broadband Electromagnetic internal condition assessment tool ... 48
3.4 Picture of CCTV internal assessment tool ... 49
3.5 Picture of Teletest tool ... 50

3.6 Schematic showing magnetic flux leakage technique 51

3.7 Picture of magnetic flux leakage external condition assessment tool.......... 52

3.8 Cross section of RFEC ILI tool 54

3.9 Picture of Pressure Pipe Inspection Company's RFEC/TC tool.......... 55

3.10 Example of inertial mapping pig from the PII Pipeline Solutions range.......... 59

3.11 High Resolution 32 Sensor caliper pig 60

3.12 BJ Pipeline Inspection Services Vectra MFL Tool.......... 60

3.13 Pictures of ultrasonic tools from the PII Pipeline Solutions range 61

3.14 Draft selection procedure for condition assessment inspection of water transmission mains.......... 64

H.1 Lord McGowan's Bridge – the original crossing 102

H.2 The successfully replaced duplicated mains 102

I.1 City of Aurora Collection System.......... 111

I.2 Initial Investigation. 113

I.3 Autopsy begins with incision.......... 113

I.4 Top of pipe shows extensive distress.......... 114

I.5 Bottom of pipe is free of distress 114

I.6 Distress area with 8 broken wires 115

I.7 Distress area with 12 broken wires 115

I.8 Typical "critical" pipeline bridge crossing 118

I.9 Pipeline asset management process 118

I.10 Life cycle of typical pipeline 120

I.11 Natural corrosion of spun gray cast iron pipe (BS 1211) 122

I.12 Imaging of inspection outputs 121
I.13 Small diameter inspection tool downloading data 123
I.14 ECAT – External Condition Assessment Tool 124
I.15 3D graphical representation of corrosion pits in iron pipe 126
I.16 S19 – External Condition Assessment Tool 127
I.17 3D graphical representation of pipeline structure 128

FOREWORD

The Awwa Research Foundation is a nonprofit corporation that is dedicated to the implementation of a research effort to help utilities respond to regulatory requirements and traditional high-priority concerns of the industry. The research agenda is developed through a process of consultation with subscribers and drinking water professionals. Under the umbrella of a Strategic Research Plan, the Research Advisory Council prioritizes the suggested projects based upon current and future needs, applicability, and past work; the recommendations are forwarded to the Board of Trustees for final selection. The foundation also sponsors research projects through the unsolicited proposal process; the Collaborative Research, Research Applications, and Tailored Collaborative programs; and various joint research efforts with organizations such as the U.S. Environmental Protection Agency, the U.S. Bureau of Reclamation, and the Association of California Water Agencies.

This publication is a result of one of those sponsored studies, and it is hoped that its findings will be applied in communities throughout the world. The following report serves not only as a means of communicating the results of the water industry's centralized research program but also as a tool to enlist the further support of the nonmember utilities and individuals.

Projects are managed closely from their inception to the final report by the foundation's staff and large cadre of volunteers who willingly contribute their time and expertise. The foundation serves a planning and management function and awards contracts to other institutions such as water utilities, universities, and engineering firms. The funding for this research effort comes primarily from the Subscription Program, though which water utilities subscribe to the research program and make an annual payment proportionate to the volume of water they deliver and consultants and manufacturers subscribe based on their annual billings. The program offers a cost-effective and fair method for funding research in the public interest.

A broad spectrum of water supply issues is addressed by the foundation's research agenda: resources, treatment and operations, distribution and storage, water quality and analysis, toxicology, economics, and management. The ultimate purpose of the coordinated effort is to assist water suppliers to provide the highest possible quality of water economically and reliably.

The true benefits are realized when the results are implemented at the utility level. The foundation's trustees are pleased to offer this publication as a contribution toward that end.

It is increasingly important that water utilities gain a better understanding of the performance and current condition of their buried assets through regular inspection and assessment in order to develop robust and cost effective capital and operational maintenance programs. This report summarizes a state-of-the-art literature review of potential non-interruptive condition assessment inspection devices for large diameter transmission mains, and the outputs from an expert workshop which reviewed the performance of current technologies and future condition assessment research needs.

Walter J. Bishop
Chair, Board of Trustees
Awwa Research Foundation

James F. Manwaring, P.E.
Executive Director
Awwa Research Foundation

ACKNOWLEDGMENTS

The authors of the report would like to thank all those who helped contribute to the project through attendance and contributions to the success of the workshop, and participation in the telephone consultation. They are listed in the appendices at the end of this report.

The support and hospitality of United Water Services is particularly acknowledged for their assistance with the hosting of the workshop at Haworth Water Treatment Plant, New Jersey.

In addition, the authors would also like to thank the members of the Project Advisory Committee (PAC)—including John Proctor, Yorkshire Water, Bradford, West Yorkshire, United Kingdom; Albert Wtorkowski, Chicago Department of Water Management, Chicago, Ill.; Mike Royer, United States Environmental Protection Agency, Edison, N.J. for their guidance and support. Thanks also go to Colin Millington, Fovarosi Vizmuvek, Budapest, Hungary, for his help in assessing the initial telephone consultation process.

The authors would also like to thank Susan Ferguson, Daniel Robson and Marie McNally, NWTC, for their support throughout this project.

EXECUTIVE SUMMARY

INTRODUCTION TO THE PROJECT

The aim of the project was to conduct a state-of-the-art literature review of non-interruptive condition assessment inspection devices for large diameter transmission mains e.g. greater than 12 inch (300 mm). In addition, an expert panel workshop was to be held to review business needs and drivers, the performance of existing technologies and buried asset condition assessment research needs.

BACKGROUND

Many water utilities have only limited knowledge of the condition of their assets. It is increasingly important that these utility companies gain a better understanding of the current condition and performance of these buried assets. Regular condition assessment of mains can greatly assist utilities with developing robust and cost effective operational maintenance programs.

The removal of pipe sections, for condition assessment, has been common practice in the water industry. This technique causes interruption to supply and results in the assessment of a small percentage of the network as a whole. Clearly, any non-interruptive condition assessment technology that is both technically and economically appropriate would offer significant opportunities and benefits to the pipeline operator.

There are a number of existing and developing condition assessment tools available to the operators of supply networks. Using these techniques, it is possible not only to assess cracks, wall thickness, internal and external metal loss associated with corrosion, but also mechanical damage and manufacturing faults. This information would enable the identification of mains requiring local repair and the prioritization of mains for replacement or rehabilitation.

As condition assessment tools and techniques become more advanced, they should help to ensure that investment in network maintenance is properly targeted. Where the consequences of pipe failure are very serious, as in the gas and petrochemical industries, much work has already taken place to develop condition assessment, inspection and risk assessment techniques.

PROJECT APPROACH

The purpose of the project was to conduct a state-of-the-art literature review of non-interruptive condition assessment inspection devices for large diameter transmission mains and then to bring together a group of experts to review condition assessment research needs and the performance of technologies. Specifically the research team conducted the following tasks:

- A detailed literature and manufacturer search of non-interruptive condition assessment inspection devices for large diameter transmission mains.
- A telephone consultation to understand the business needs, the business drivers, and the current use of condition assessment technologies.
- Identification of experts to participate in the planned project workshop.

- A 3-day workshop to review business needs, the performance of existing technologies, and asset condition assessment research needs.
- A project web site was established to promote the project and to share information on the project.
- Reporting on the findings of the literature survey, consultation, and conclusions from the workshop.

RESULTS

The Business Needs Review considered some of the current issues facing the Water Industry network managers and the potential for pipeline condition assessment techniques to assist in meeting some of these needs.

There are a number of high level drivers in the Water Industry, which could influence the adoption of inspection devices. These include legislative requirements, levels of service, safety issues, social costs and loss of fire fighting water supplies.

The target for utility asset management is to meet a required level of service in the most cost-effective way. To achieve this, utilities need to have a clear understanding of their assets and how they are performing. Condition assessment inspection devices can assist with this process and help the decision-making process for investment. Either deferral or reduction of capital expenditure and a reduction in operating expenditure can justify business cases.

A telephone consultation with 41 water utilities provided a clearer understanding of the size and material composition of water transmission networks, the perceived problems associated with them, the level of understanding of networks, the current methods of assessment, and the use, appeal and needs for non-interruptive condition assessment inspection devices.

A review was also undertaken of the drivers in other pipeline industries and the application of condition assessment.

The Technological Review, which built upon previous work undertaken by AwwaRF, has identified and reviewed the current available and developing technologies. A number of technologies, including Acoustic Emission Monitoring, Acoustic Leak Detection, Broadband Electromagnetic, CCTV, Long Range Ultrasonic (Guided Wave), Magnetic Flux Leakage, Remote Field Eddy Current, Remote Field Eddy Current – Transformer Coupling, Seismic Pulse Echo and Ultrasonic are available to assist with condition assessment. In addition, there are a number of traditional technologies, such as pipeline history, location of mains, pressure testing, flow tests and Hazen Williams C tests, soil resistivity, and coupons and samples.

A successful 3-day workshop was held as one of the key elements of the project. This enabled 49 invited experts to review the identified business needs and application of available technologies and contribute to this report.

A Road Map has been developed to assist network managers in selecting the most appropriate available technology.

A key element of the workshop was the forum for the experts to assist the project team to identify key parameters and their importance for different materials. A generic specification has been developed and the existing technologies assessed against the parameters for the three main groups of materials.

CONCLUSIONS

- The technology review undertaken within this project was able to build upon previous work carried out by AwwaRF. The outputs presented are, however, more comprehensive than previous studies as they identify additional available technologies being utilized and new products and services under development.
- The study highlighted the importance of understanding the business needs of water utilities and provides an overview of some of the key business drivers and justifications.
- It is important to understand how the available technologies can assist in the decision-making process for water utilities. The application of these tools and the results obtained can contribute to the asset management process. However, there are varying levels of application and strategies existing.
- Although the commodities and the business drivers are generally different within the oil and gas industries, there are some links with the Water Industry. This is currently particularly applicable in the area of risk management of the steel assets.
- One of the strengths of this project was the ability to consider the subject on a global basis.
- The workshop provided an ideal forum to expand the understanding of the business drivers for the Water Industry and the requirements of the utilities for condition assessment. The strong panel of experts assembled from water utilities, technology providers, consultants, researchers, academia, regulators and government departments were also able to review the application of existing technologies and contribute to the ideal parameters for a non-interruptive condition assessment inspection device.
- A number of traditional technologies are being successfully used by some water utilities to assist them with their decision making process. The application of these traditional methods could fit further into decision-making processes.
- The experts attending the workshop contributed to the development of a generic parameter portfolio for an ideal specification. No single device exists to meet the requirements of this generic specification or all of the existing materials in use.
- A number of the existing technologies go some way to address some of the generic requirements. Not all of the above technologies are truly non-interruptive condition assessment inspection devices.
- It was not possible within the scope of this project to rank the potential varying assessment parameters.
- The adoption of these technologies within the Water Industry seems to be limited. Although in certain cases, the PCCP assessment technologies appear to be providing solutions for customers. Some elements which may be affecting the adoption of the technologies on a more widespread basis could be
  - Charging arrangements for service
  - Performance of technologies
  - Expectation/capability of results

RECOMMENDATIONS

- Water Utilities clearly need to define their business needs and adopt strategies to meet their individual operational circumstances. A good understanding of data management needs is essential.
- Technology providers need to review their methods of charging for services to closer suit the business needs of the Water Utilities.
- No single non-interruptive condition assessment inspection device exists for all water transmission mains. Water utilities should consider each job on its merits and select from the most appropriate technology available to provide the required results
    - Traditional technologies
    - Existing condition assessment technologies
- It is recommended that further research is undertaken to take full advantage of the outputs from this project, and facilitate the transfer of knowledge in this important area of work
    - Additional consultation would be desirable with water utilities, technology providers and the academia.
    - Enhance and maintain the project web site as a condition assessment forum.
    - Facilitate on-going workshops dedicated to condition assessment.
    - Develop a detailed specification for a non-interruptive condition assessment inspection device by material.
    - Develop a "Best Practice Selection Manual for the Condition Assessment of Water Mains".

CHAPTER 1
INTRODUCTION

BACKGROUND

Many water utilities have only limited knowledge of the condition of their assets. It is increasingly important that these utility companies gain a better understanding of the current condition and performance of these buried assets. Regular inspection and assessment of mains can greatly assist utilities with developing robust and cost effective operational maintenance programs.

The removal of pipe sections, for condition assessment, has been common practice in the water industry. This technique causes interruption to supply and results in the assessment of a small percentage as a whole. Clearly, any potential non-interruptive condition assessment technology that is both technically and economically appropriate would offer significant opportunities and benefits to the pipeline operator.

There are a number of existing and developing condition assessment tools available to the operators of supply networks. Using these techniques, it is possible to not only assess cracks, wall thickness, internal and external metal loss associated with corrosion, but also mechanical damage and manufacturing faults. This information would enable the identification of mains requiring local repair and the prioritization of mains for replacement or rehabilitation.

As condition assessment tools and techniques become more advanced they should help to ensure that investment in network maintenance is properly targeted. Where the consequences of pipe failure are very serious in the gas and petrochemical industries much work has already taken place to develop condition assessment, inspection and risk assessment techniques.

OBJECTIVES

The objective of this project was to conduct a state-of-the-art literature review of non-interruptive condition assessment inspection devices for large diameter transmission mains (greater than 12 inches). In addition, an expert panel workshop was to be held to review buried asset condition assessment research needs and performance of existing technologies.

- The key elements to be addressed under the project were as follows:
- Identify specific water purveyor needs and benefits from an effective non-interruptive inspection device
- Identify different needs, if any, of pipe material(s) predominantly used in the industry
- Identify limitations or constraints with systems and scientific knowledge currently being used in state-of-the-art condition assessment techniques
- Analyze data currently available from condition technologies (e.g. Can the data produced provide a reasonably accurate estimate of a pipe's condition or the residual service life of transmission mains?)
- Identify alternatives to non-interruptive condition inspection devices
- Identify basic design parameters for technologies (e.g. mobility, pipe size, power requirements, excavation needs, etc.)

- Identify if non-interruptive inspection devices, based on the industries needs, are feasible, and, if so, develop general guidelines for device(s) capable of in-service condition assessment, as well as cost and performance guidelines for future system development

PROJECT APPROACH

The purpose of this project was to conduct a state-of-the-art literature review of non-interruptive condition assessment inspection devices for large diameter transmission mains and then bring together a group of experts to review condition assessment research needs and the performance of technologies. Specifically the research team conducted the following tasks:

- A detailed literature and manufacturer search of non-interruptive condition assessment inspection devices for large diameter transmission mains. This search also included a review of technologies within utilities. The literature review did not simply focus on water industry devices, but also considered other technologies developed in the oil, gas and nuclear industries.
- A telephone consultation was undertaken with some of the key water utilities within the U.S.A., Canada and the U.K. to try and understand the business needs, drivers and the current use of condition assessment technologies within their operations.
- Experts, from the water utilities, technology providers and other industries with interests in condition assessment, were identified to participate in the planned project workshop. The project team, working with the Project Advisory Committee, utilized their wide network of potential contacts in this technology area.
- A 3-day workshop was held to review business needs, the performance of existing technologies and asset condition assessment research needs. The key focus of the workshop was to address the detailed project objectives.
- A project web site was established to promote the project and enable the project team to share information securely on project details.
- Reporting on the findings of the literature survey, consultation and conclusions from the workshop. Two key documents were produced - a pre-workshop synthesis report, describing identified business needs and technologies available in the water industry, and this final project report.

CONSULTATION

To obtain a better understanding of the business needs and drivers, and the use of condition assessment technologies within the Water Industry, the project team developed an approach to consult water utilities by telephone. The methodology for the telephone consultation was initially tested with Fovarosi Vizmuvek (Budapest Waterworks, Hungary) who had existing experience with condition assessment.

The target audience for the consultation was primarily the U.S.A. and Canada, with some comparison data being obtained from the U.K. Forty-one water utilities within Canada, U.K. and U.S.A. were consulted and details of the organizations involved are shown in Appendix A. A general advisory note on the project was circulated to an additional 47 water utilities in the

U.S.A. to obtain any additional interest. Details of the additional utilities notified are shown in Appendix B.

The consultation was geared at better understanding the size and material composition of water transmission networks, and the perceived problems associated with them. Utilities were consulted to determine their understanding of their networks, the current methods of assessment, and the use, appeal and needs for non-interruptive condition assessment inspection devices.

Details of the results of the consultation are presented in the various figures within Chapter 2. An attempt has been made to compare the U.K. market to the markets in the U.S.A. and Canada.

WORKSHOP

A key element to this project was the expert panel workshop, which was held at Haworth Water Treatment Plant, New Jersey, U.S.A. with the support of United Water Services. The workshop was held between the 12th and 14th March 2003 to review business needs and drivers, the application of existing condition assessment technologies and future specification needs.

One of the strengths of the project team was their ability to identify and assemble a diverse group of experts and bring an international perspective to the project outputs. It was important to include a core of water utilities at the workshop and details of the organizations invited are shown in Appendix C. The water utilities were supplemented with representatives from the technology providers, other pipeline industries, consultants, researchers, government bodies and regulators associated with the subject, and some of the project team. Forty-nine delegates attended the workshop and details of the participants are shown in Appendix D.

Delegates attending were provided with a pre-synthesis report from the project team. This report included the initial results from the technology review and identified some of the perceived business needs and drivers, and the potential justification for the use of non-interruptive condition assessment inspection devices. The document was specifically geared towards encouraging knowledge sharing and discussion during the workshop.

The workshop was split between plenary sessions, and work group forums and reporting. Key speakers were identified to reinforce the pre-workshop synthesis report with presentations from water utilities, technology providers, and other pipeline industries. Three work group forums were held to address the detailed objectives of the project, and considered "The Water Industry Needs", "The Suitability of Existing Technologies," and "General Guidelines for Technology". Delegates were also provided with a workshop pack containing copies of all of the presentations and summaries provided by each of the technology providers.

The workshop proved to be very successful and has contributed well to the production of this report. The event was generally well received by the delegates attending with their feedback being detailed in Appendices E to G. Delegates were requested to rank their feedback from 1 to 5 with 5 being the highest ranking. The average feedback scores from the Water Utilities, Technology Providers, Consultants and Academia were 4.14, 2.92 and 4.25, respectively.

WEB SITE

A web site was created for the duration of the project. This facility was provided to promote the project and publish information on events. A secure area was provided for the project team to share information during the life of the project.

CHAPTER 2
BUSINESS NEEDS REVIEW

INTRODUCTION

This chapter considers some of the current issues facing the Water Industry network managers and the potential for pipeline condition assessment techniques to assist in meeting some of these needs. Many underground assets are old and are failing, leading to bursts, leaks and consequential damage. Condition assessment can assist with developing an understanding of the extent of these network problems and can also be used to better target the investment required by the Water Industry in order to maintain water supply and quality to customers into the 21st Century.

Initial consideration is given to the business drivers that may encourage the Water Industry to make greater use of condition assessment techniques. To achieve a wider perspective, consideration is also given to the use of condition assessment by the other pipeline industries, as well as by water companies in North America and Europe. In order to try and quantify the potential benefits of undertaking a structured condition assessment program, an attempt is also made to offer a number of factors that may be considered in any business justification process. This is supported by a number of case studies highlighting work already undertaken by water utilities. Chapter 4 considers the industry requirements for any potential condition assessment tool, and any subsequent operational constraints that would need to be overcome.

BUSINESS DRIVERS

This section examines the global pipe condition assessment needs of the water and other industries and what external forces are making water utilities consider undertaking this process. This section is split into three subsections. The first considers condition assessment within the oil, gas and chemical industries, why assessment of pipe condition is undertaken, the risks associated with loss of containment, and legislation. The second investigates the high level drivers for condition assessment within the water industry such as legislation, maintaining levels of service to customers, safety, social costs and loss of fire fighting supply. The third subsection considers the role of condition assessment within asset management. Topics include assessment of structural condition of the network, leakage reduction, the improvement of rehabilitation decision making, predicting future failures, reduction of operational expenditure, maintaining security of supply and risk assessment planning. Case studies are presented which demonstrate some of the drivers outlined throughout this section of the report.

Condition Assessment in Other Industries

Non Water Industry Drivers for Non-Interruptive Condition Assessment

The world's oil and gas transmission system is extensive. In Western Europe alone it covers many thousands of kilometers in length, and worldwide there are millions of kilometers. With ever-increasing pressure on oil and gas pipeline operators to continuously monitor and measure their pipeline systems, 'in-pipe' inspections have become an integral part of risk management and integrity verification programmes.

Due to the nature of their business, the chemical and petrochemical industries already extensively use non-destructive testing of their pipes, and much effort is put into maintaining and enhancing natural gas system reliability and integrity (Anderson 2003). Many of these pipelines are designed with internal inspection in mind, and some have inspection-pig launch and retrieval systems integrated into their pipelines at the time of installation. Most commonly, non-destructive evaluation techniques are employed across these industries. New techniques have also been embraced; for example, laser profilometry is starting to be widely used in the chemical, petrochemical and in military organizations (Roberts 1999). These industries recognize the value of inspecting their assets as it saves them time, money, loss of production and potentially the lives of employees and the surrounding community.

Inspection and maintenance are crucial to the economic survival of a petrochemical company. Without them, a company would find its position untenable, as it would be constantly suffering loss of production, paying for cleanup and claims for compensation. A CEO of a petrochemical company said, "Maintenance? That's my 'stay-in-business' capital. I have no choice" (Trench and Sturbitts 1999).

In the U.S., the existing Office of Pipeline Safety regulations requires certain periodic pipeline integrity tests, such as checks on valve operation and cathodic protection against corrosion. In recent years, the emerging technologies for in-line inspection have matured rapidly. Pipeline companies have been guiding these developments through research programs and by working with non-destructive evaluation suppliers. Based on the results of these efforts, the oil pipeline industry believed it was time for the Office of Pipeline Safety to promote more extensive use of in-line pipeline test technology. The Office of Pipeline Safety already had the statutory authority to impose more stringent and comprehensive pipeline integrity testing requirements. It was generally agreed that increased testing, particularly in areas where communities have grown up around pipelines and other high consequence areas, was favorable (AOPL 2000). As a result, the DoT Pipeline Integrity Management Regulations enacted at the end of year 2000, specifically permitted the use of in-line inspection technology.

Pipe condition assessment is well established in other industries. For example, an ultrasonic testing device created by a British company is used in the state-run nuclear power research institute in China to test the status of the welds on their piping. Radiographic wall thickness testing is used in the chemical and petrochemical industries (Hecht, Bauer, and Lindemeier 1998). In the chemical industry, eddy current and ultrasonic non-destructive testing techniques are used to inspect heat exchanger tubes (Rasmussen, Kristensen, and Jeppesen 1998). Another form of eddy current technology, called pulsed eddy current, is used in the chemical and marine inspection industries to inspect coated and insulated pipes and other surfaces, where removal of coatings or insulation would be expensive or cause problems (Robers and Scottini 2002). Some companies use a continuous assessment to constantly monitor the status of their pipe network. An ultrasonic system permanently fixed to a pipe network has been successfully field tested in the petrochemical industry in all weather conditions for over a year (Deutsch, Platte, Schäfer and Müller 2002).

The decision to operate an inspection device within a pipeline may be driven by a number of factors: the primary drivers being fitness for purpose assessment, safeguarding assets and optimizing efficiency, or complying with regulations regarding safety and the environment. Should a pipeline fail, the consequences may impact upon safety, the environment, legislation, public relations, production and economic costs. Management of these aging assets is now a high priority with corrosion and cracks the major concern (Kopp 2003).

The management of pipeline systems requires a dedicated pipeline failure management strategy that understands the risks involved. Risk modeling is undertaken to manage the likelihood and consequences of failure at an acceptable level (Wayman 2003). The cost of pipeline failure is of such magnitude that preventive measures, predictive methodology and defect elimination should be used as enabling techniques to achieve the maximum pipeline performance, this is where the use of in-pipe inspection tools is extremely valuable. High quality data acquisition on the current and future predicted condition of pipeline systems is now possible and the interpretation of derived risk data provides a great deal of insight to pipeline condition. Performance of pipeline systems can be monitored over the long term by measuring the risk of deterioration to the pipeline components (Crocket and Maguire 1999).

The main drivers for in-pipe inspection tools will be discussed more fully in the following sections.

Business and Economic Factors

Pipelines are vital arteries for energy flow. The value of the hydrocarbons flowing through them is significant in terms of national economies. For example, a 36-inch (914 mm) diameter oil line can carry as much as 1 million barrels of oil per day from the Forties oilfield located 105 miles (170 kilometers) northeast of Aberdeen, Scotland. At $25 per barrel, such a pipeline can deliver $25 million worth of oil per day or $9 billion worth annually. Gas pipelines too are vital arteries; a 40-inch (1,016 mm) diameter pipeline such as the Inter-connector from the U.K. to Zeebrugge in Belgium is designed to handle up to 20 billion cubic metres of gas per annum. This is equal to one third of the total U.K. gas consumption or twice that of Belgium. At U.K. domestic prices, this is equivalent to more than $4 billion worth of gas annually (PII Pipeline Solutions 2002).

It is vital to establish a pipeline maintenance and inspection strategy that ensures transportation and delivery of the product to the satisfaction of the shipper and/or customer. It must be robustly planned to optimize performance by increasing the overall life by ensuring that the probability of failure remains at an acceptable level whilst minimizing overall operating costs (Fordyce and Kidd 1999).

A strategy of periodic internal inspection followed by integrity assessments using fitness for purpose criteria and selective remedial action, is the most cost beneficial way of extending the performance of aging pipelines (Jones 2002). This would include extending the operational life of pipelines.

What Can the Water Industry Learn from Other Industries?

The concept of a risk-based approach to asset management could be applicable to pipelines in the water industry. There are however, several fundamental differences between the high-pressure systems operated by oil and gas utilities and the lower pressure systems operated by water companies. These are summarized in Table 2.1 (Wayman 2003).

Table 2.1

High and low pressure pipelines

High pressure pipelines	Low pressure pipelines
High value	Perceived low value
Major safety issues	Variable construction costs
Good design standards	Variable criticality
High quality steels	Design variability
Good corrosion protection	Materials
Regular maintenance	Corrosion protection
Defect criticality understood	Maintenance deficiencies

Source: Wayman, M., Risk Assessment in the Pipeline Industry, 2003.

Risk modeling for low pressure pipelines can be complex, due to wide variations in original design and construction standards, pipe materials, loading regimes, etc, and changes to these parameters over time. The main requirement is to determine the probability of pipe failure and the likely consequences, and compare them in a mathematically robust way. Nevertheless, the fundamental processes involved in generating risk models would be applicable as part of the asset management of critical mains (which would need to be identified within the network). The process can be summarized as shown in Figure 2.1.

For large diameter pipes, it is necessary for the structurally important defects to be defined as part of the condition assessment process. For high-pressure pipelines, effectively working as pressure vessels, defect sizes considered to be important are well defined, based upon fracture mechanics. However, for large diameter low pressure pipelines, a range of defects can be considered critical, depending on internal pressure, external loading regime and the way in which the pipeline is supported. The scale of deterioration, which potentially would lead to failure, will also vary with pipe diameter, wall thickness, loading regime etc. The condition assessment process must therefore consider the criticality of a range of defects and focus inspections towards the most critical. A major influence on pipe deterioration is the type and quality of corrosion protection originally applied to the pipe and how it has been maintained during its service life.

A simplified version of this approach has already been undertaken by Northumbrian Water Limited (Morrow 2003) in the management of their strategic crossings (points within the distribution system where single supply mains cross rivers, trunk roads or railway lines). The aim of the program is to determine and understand the level of risk posed at each of these crossings in a systematic and controlled way to better manage the level of risk of loss of supply to customers. The approach involves identifying critically loaded sections and utilizing appropriate inspection techniques as part of an overall condition assessment process.

WRc have also expanded their approach from purely pipe condition based decision making on trunk mains to an approach based more on a broader assessment of pipeline risk (Conroy 2003). Risk assessment software tools have also now emerged which consider a broader range of factors. In the risk based decision making process, the pipeline owner can establish what is an acceptable level of risk, and then pipelines which exceed this now or in the future should be assessed in more detail and risk mitigation options evaluated.

The National Research Council (NRC) in Canada have also developed a risk based approach based upon inspection, condition assessment, assessing the risk of failure and then

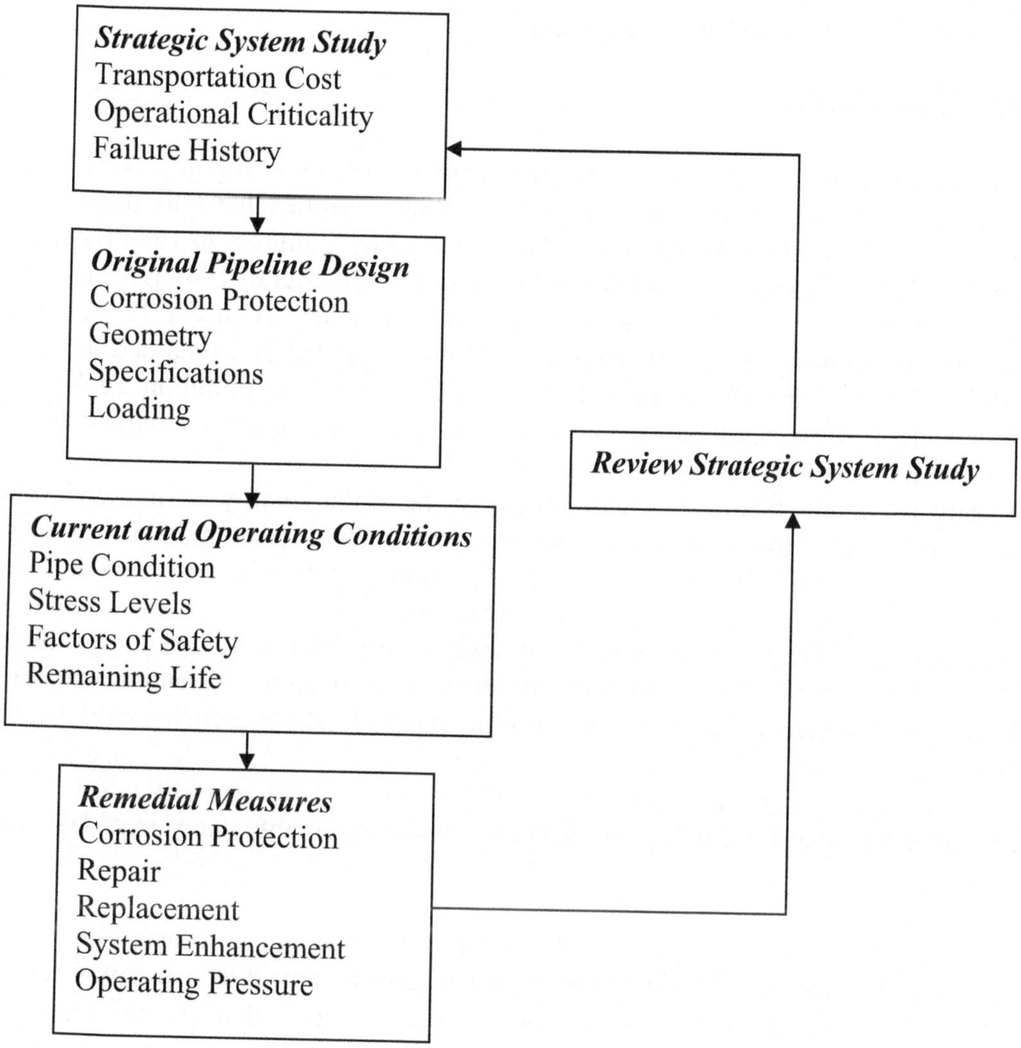

Figure 2.1 Risk model (*Source*: Wayman, M., Risk Assessment in the Pipeline Industry, 2003)

deciding upon the optimum (most economic) time for renewal of the asset (Rajani 2000). This approach can be applied to various pipe materials including PCCP, gray and ductile cast iron.

Other parameters such as water quality and hydraulic performance can also be considered within the analysis.

It is clear therefore, that the adoption of any risk-based approach to the management of pipeline assets will also need to include an element of pipe condition assessment within the process.

High Level Drivers within the Water Industry

Legislative Requirements

In the U.S., the United States Environmental Protection Agency (USEPA) regulates health related issues within the water industry. Water utilities have to meet the standards pursuant to the Safe Drinking Water Act (SDWA) passed by Congress in 1974 (later amended in 1986, 1996 and 2002) and the Federal Water Pollution Control Act of 1948 (Clean Water Act). A summary is given in "25 years of the safe drinking water act" (USEPA 1999). The Federal Water Pollution Control Act was enacted in 1972 and amended and renamed the Clean Water Act in 1977, and finally re-authorised in 1991. The CWA regulates pollution discharges to water bodies. Together, these regulations put strict limits on water quality supplied to homes and businesses by public and privately owned utilities.

There are many factors that can influence the water quality as it travels through the distribution networks. Infiltration and permeation of external contaminants can affect water quality, as can leaching of particles from old pipe lining materials. Also in water distribution networks, corrosion of the carrier pipes can seriously affect the quality of water being supplied to customers as it can increase levels of turbidity above prescribed limits leading to taste and odor problems. Tuburculation from corrosion products can become a habitat for biofilms or pathogens, and can effect the levels of chlorine residual, all of which could lead to health problems.

Public water systems defined in Code of Federal Regulations (CFR) 40 CFR Part 141.2 are those that provide at least fifteen connections or supply potable water to at least twenty-five people.

- *Community water systems:* public water systems that serve at least fifteen connections used by year-round residents or regularly serve twenty-five year-round residents.
- *Non-community water systems:* public water systems that are not community water systems. A non-community water system is either a "transient non-community water system (TWS) of a non-transient non-community water system (NTNCWS).
- *Non-transient non-community water systems:* public water systems that are not community water systems but that serve at least the same twenty-five people for 6 months per year.
- *Transient non-community water systems:* a non-community water system that does not regularly serve at least twenty-five of the same persons over six months per year.

There are other U.S. legislative drivers which may lead to an increased need for the use of condition assessment techniques. The existing SDWA mandated Drinking Water State Revolving Fund (DWSRF) is managed by the USEPA. The distribution of DWSRF funds among and within states is based upon a demonstration of need, and inspection could help to demonstrate legitimate need. The 1996 SDWA Amendments directed the USEPA to conduct a survey of infrastructure needs facing public water systems, the first report was published in 1997. The second survey released in 2001 estimated that drinking water systems would need to invest $150.9 billion over a 20-year period to ensure the continued provision of safe drinking water. Results from the survey are used to develop formula to allocate DWSRF grants (Drinking Water Needs Survey, 2001).

To obtain a better understanding of the challenges the clean and drinking water industries face with aging infrastructure, the USEPA has conducted a study "The Clean Water and Drinking Water Infrastructure Gap Analysis" to identify whether a funding gap will develop between projected investment needs and projected spending. Improved asset management is regarded as a potential approach to helping ensure sustainable water infrastructure, and condition assessment is an important component of asset management.

A further issue, which may potentially become a driver for the use of condition assessment techniques by water utilities in the United States, is GASB34. GASB is the Governmental Accounting Standards Board, which monitors the financial reporting methods for state and local governments. 34 refers to Statement No. 34: Basic Financial Statements – and Management's Discussion and Analysis – for State and Local Governments, in effect since 1979. In June 1999, the GASB proposed significant changes to Statement No. 34. The new initiative known as GASB34 intended to better reflect the way the current infrastructure is accounted for, thereby making financial reporting easier and more comprehensive for today's state and local government agencies. The new GASB34 standards provide two methods for reporting their infrastructure assets:

- Reporting of infrastructure assets is calculated by the historic cost less depreciation
- Reporting is done by the infrastructure asset management system, also known as the modified approach, and is based upon the agency meeting the following requirements:
    - Have an up-to-date inventory of eligible infrastructure assets
    - Perform condition assessments of the eligible infrastructure assets every three years and summarize the results using a measurement scale
    - Estimate each year the annual amount to maintain and preserve the eligible infrastructure assets at the condition level established and disclosed by the government

Although the feeling in the workshop discussion groups was that this initiative would not yet be a major driver for condition assessment activities, it could encourage a number of water utilities to consider condition assessment techniques to help with their asset reporting requirements.

A further consideration in the use of condition assessment techniques to help better target investment, could be the increased activities under the Homeland Security initiative, which may highlight that overall funds for infrastructure renewal may be limited, and may not be sufficient to meet the utilities requirements.

In the U.K., the water industry is regulated by the Department for Environment, Food and Rural Affairs (DEFRA), the Office of Water Services (OFWAT) and the Drinking Water Inspectorate (DWI). DEFRA tables legislation to be put forward to parliament and the Secretary of State has ultimate authority over the water industry. OFWAT has the task of regulating the water industry and ensuring that water companies adhere to agreed levels of service, and can fine water companies if they fail to do so. The Drinking Water Inspectorate (DWI) assesses the quality of drinking water in England and Wales and takes action if the standards are not being met. Under the Water Industry Act 1991, a water utility must be able to maintain a supply of water to premises and maintain the connection between a water main and a service pipe. If this service is disconnected without any reasonable steps being taken to prevent it, a water company will be liable to prosecution. The water utility must also supply water that is deemed "fit for human consumption." Failure to meet this requirement can in theory lead to a custodial sentence and a fine if convicted (although this has never yet actually occurred within the UK water industry). At the request of a business or factory owner, a water utility is obliged to install a fire hydrant for fire fighting purposes only. A water company also has to install fire hydrants on its water mains (other than trunk mains) wherever the fire authority requests they should be installed. It is also the duty of a water company (under the Water Industry Act 1991) to ensure that fire hydrants are in good working order.

It is important to maintain high standards of service and water quality to meet the requirements set by government. Condition assessment can help in this regard as it can be used to monitor the deterioration of pipes and to help with predicting when and where pipes are likely to fail and identify where corrosion may be compromising water quality.

Levels of Service (LOS)

The levels of service are the defined quality for a particular activity against which service performance can be measured. Service levels usually relate to quality, quantity, reliability, responsiveness, environmental acceptability and cost (INGENIUM and IPWEA 2002). Examples could be the response times of the utility to leaks or breaks, network pressure and flow at connections and numbers of customer complaints. Levels of service are significant performance targets for most water companies and form the minimum operational standards in terms of the provision of a safe, continuous water supply and appropriate contact with customers. The levels of service requirements can vary from country to country and even from contract to contract, but the key elements are generally the same. The LOS reflect the strategic objectives of the utility and are usually based on

- Customer expectations
- Legislative requirements
- Utility mission and objectives
- Availability of resources
- Financial constraints
- Utility delivery mechanisms

It is important that the asset owners measure the current LOS that their assets deliver, their customers expected LOS, and the customer's ability or willingness to pay for that LOS. The asset owner can then manage the LOS to match the expectations of the customers and their ability to pay.

As well as internal LOS goals, many countries now have a formal body responsible for monitoring the performance of water companies against the required levels of service whether it is the local government, State or Federal bodies. In the U.K. it is OFWAT, with the agencies for drinking water quality and treatment of wastewater being the DWI and Environmental Agency (EA), respectively. In the United States, the USEPA fulfills some of these functions.

In many parts of the world, the failure to achieve defined standards can be significant for a water company. For example in the U.K., under the Water Industry Act 1991, as well as meeting all of the water quality requirements, a water company is required to maintain sufficient water pressure to reach the top storey of every building within its supply area. However, a water utility does not have to supply water to a height greater than it will flow by gravitation from the supply tank or water reservoir. Levels of service often include a minimum reference level for pressure in water mains when demand is not abnormal. Companies may be required to maintain registers, which identify the properties at risk of receiving low pressure. Companies may also need to report on the numbers of properties experiencing interruptions to their supply for various periods of time, which are the responsibility of the Water Company, but are not planned or warned. Again interruptions due to bursts would fall into this category.

Pipe bursts, leakage, corrosion or tuberculation will lead to a loss of pressure and may cause complaints from customers. Inspection of a pipe network can reduce the number of complaints received by a water company as potential problem pipes can be identified and can be repaired or replaced before such inconveniences arise.

In some utility contracts, there may be agreed financial penalties for failing to meet the required levels of service. In Scotland for example, any water company has to pay compensation to both domestic and non-domestic customers if they are inconvenienced by loss of service by the order of the Water Industry Commissioner for Scotland. Compensation can be as much as £100 per domestic customer and up to £5,000 per commercial customer (Sutherland 2000).

It can be seen, therefore, that any interruptions to supply in areas fed by large diameter mains could be extremely expensive if compensation payments are required. Irrespective of the economics, any regular breach of levels of service, or inconvenience to customers may lead in the longer term to poor public relations, or even worse, to the renegotiation or loss of the contract and therefore needs to be minimized.

Safety Issues

In other fluid transport industries, particularly the petrochemical, gas and nuclear industries, there already is a culture of safety and environmental protection, as the results of pipeline failure can be catastrophic. There have been incidents that have caused environmental damage by loss of containment (Trench and Sturbitts 1999). Although escaping water is not as damaging to the environment as say oil, it can damage soil structure and cause damage to property. Water companies must meet the standards of safety; water quality and service set down by governments and regulatory bodies around the world. These new public concerns over safety and the environment, and legislation from governments have shifted the emphasis of running a utility away from financial aspects and more towards the engineering and environmental outlook (Vollman 1998).

It may be difficult to attribute costs to the impact on the environment of water pipeline failures, but an approximation can be made by using the clean up costs, both those immediately incurred and any longer term affects depending upon the sensitivity of the local habitat (Morgan and Heath 1999). The financial costs of an incident can be estimated using a spreadsheet developed for a previous AwwaRF project called the 'Grand Central Model'. The spreadsheet can estimate the cost of a single incident, the 'unit' cost of a series of similar incidents, or the cost of a series of incident over a given period of time. The last model can be useful for planning a long-term rehabilitation strategy (Cromwell et al. 2002).

These may be direct and indirect, and extend well beyond the immediate time frame of the incident. Companies incur costs at the time of an incident such as direct costs for response, clean up and regulatory reporting, as well as the costs for management time. Furthermore, the company loses revenue while the system is disrupted and then come the costs to analyse the incident and evaluate whether corrective procedures are needed. Next, there may be additional outlays for fines or penalties, and the possibility of legal liabilities and judgements or Natural Resource Damage Assessments, all of which may take years to adjudicate. Finally, the bad will and damage to the company's reputation may impede its operation for an incalculable amount of time, and thus impose additional, possibly catastrophic, costs (Trench and Sturbitts 1999).

Effective maintenance and inspection is essential to minimize the risks to safety and the environment caused by failure of items of plant. It is also essential to ensure maintenance and inspection activities minimize impact on the safety of the public, staff and contractors or on the local environment (Fordyce and Kidd 1999).

Social Costs

The loss of water supply due to pipe failure can cost a water company not only financially, but also socially. By law, a water company must ensure that a supply of water is available to customers. If the water supply is interrupted for an extended period, the water utility must ensure that water is available i.e. standpipes, bottled water or water trucks. Failure to meet these requirements can incur fines from regulatory authorities. Indirect costs, for example those associated with lane rentals paid to municipalities for work in the highway, can compound the financial effects such as legal costs and compensation claims from customers. These claims can also be raised for damage to property, which can run into millions of U.S. dollars for a major leak. A loss of supply can also generate negative publicity for a water company, caused by traffic disruption, loss of consumer confidence, loss of fire fighting supplies and ultimately have a detrimental effect on a water company's share price as profits are cut. There is also the risk of contamination of water supply from the surrounding soil, an increased risk of back-flow and incursion of debris into corroded pipes leading to water quality problems (Makar and Kleiner 2000).

Loss of Fire Fighting Water Supply

The loss of fire fighting water supply can have severe consequences, and can put both life and property at risk. Fire crews that have insufficient water mains pressure or flow would be unable to perform their duties effectively. Again pipe inspection would help avoid this by ensuring that a water company knows that the pipes supplying fire hydrants are in good condition. If they are not fit for purpose, then the appropriate action can be planned.

For example in the United Kingdom, any water company could be liable for prosecution if a loss of water pressure for fire fighting occurs under Section 65, Subsections (1), (10) and (11) of the Water Industry Act 1991. A water company must have "taken all reasonable steps and exercised all due diligence to avoid the commission of the offence"; not inspecting a water company's pipes (as potential tools are now available) could be interpreted as not exercising all due diligence in this matter.

Condition Assessment and Asset Management

Asset Management

The goal of utility asset management is to meet a required level of service in the most cost-effective way through the creation, acquisition, maintenance, operation, rehabilitation and disposal of assets to provide for present and future customers (INGENIUM and IPWEA 2002). Key elements in any generic asset management process would involve the assessment of the operating performance of the system including condition analysis, and then to continue to monitor the performance and either manage the associated risk or take remedial action when necessary. Within this key component, condition assessment is an integral part.

There are a number of drivers for improving asset management (INGENIUM and IPWEA 2002) including the following:

- In many cases, water utility networks are thought to be nearing the end of their design and economic life.
- Apart from localized pipe sampling, little system analysis has been done to gain reliable information about the condition and performance of below ground assets to support maintenance and renewal decisions.
- Environmental regulators are applying stricter criteria.
- Financial regulators are insisting that local governments recognize and equitably recover the full costs of owning and operating infrastructure over the whole life of the assets, and that both public and private sector network operators fully justify their capital and operations expenditure programs and related price structures.
- Failure of networks to deliver services through inadequate risk management or poor planning.
- Increased customer expectations resulting from greater awareness of regulatory requirements, consumer rights and reasons for previous service failures.
- New technology is providing asset owners with increased capability to monitor, evaluate and optimize asset management activities and expenditure.

Utilities therefore need to have a clear understanding of their assets and how they are performing. All management decisions regarding maintenance, rehabilitation and renewal revolve around these aspects. A limited understanding of the current condition or performance level of an asset may lead to premature failure of that asset and leaves the utility with only one option which is to replace the asset (generally the most expensive option). The unforeseen failure of an asset can have consequences that constitute a business risk or potential loss to the utility.

There are many different condition assessment techniques that can be applied to different situations (pipe material, diameter, geometry etc.). One of the most common techniques is remote field eddy current (RFEC) which generates magnetic fields in ferrous pipes and detects faults in a pipe by detecting changes in the uniformity of the magnetic field. Another common technique is ultrasonic; this sends pulses of ultra-high frequency sound waves into a pipe wall and detects defects by comparing the reflected sound wave to the original. These and other techniques (and their variations) will be discussed in more detail in the Technological Review in Chapter 3.

By conducting regular condition and performance monitoring exercises, maintenance and rehabilitation strategies can be updated and refined, and replacement programs determined with more accuracy. Even if it is found that a failure is imminent, the utility will still have time to consider other options, or in the event of failure will be able to reduce some of the consequences. Condition assessment will allow the utility to better understand the remaining life of its assets, which drives future expenditure patterns.

On average, relining of a pipe is only 43% of the cost of replacement and the cost benefits of relining increase proportionally with an increase in pipe diameter (Sutton and East Surrey Water Plc. 2000). The benefits of rehabilitation of large diameter transmission mains instead of replacement have been shown. The cost of relining a pipe with a semi-structural or structural lining can be recouped within 10 to 20 years from savings in lost water and structural repair costs (Heavens 2003).

For many water companies, this would be the most important business driver in the use of condition assessment techniques, and during the workshop, attendees ranked this number 1 overall. The benefits of condition assessment within the overall asset management process were agreed by the attendees to be as follows in order of preference:

- Rehabilitation decision making
- Future failure prediction
- Leakage reduction
- Risk management
- Asset valuation
- Reduction in OPEX

Rehabilitation Decision Making

There are many factors, which help the network manager to identify mains for future rehabilitation or replacement and these would include burst history, consequences of failure, water quality problems, leakage and poor hydraulic performance. However, without a more complete understanding of the overall condition of the main in question, it is often more difficult to decide on the most appropriate remedial action. This can be to do nothing and manage the risk, to clean affected sections, to repair damaged sections only, to rehabilitate or replace damaged sections only, to rehabilitate the whole pipe length or to replace the whole pipe length. This decision is often based upon the potential consequences of failure, previous experience, regulation, local politics, funding available etc, and sometimes, limited local condition assessment is performed to increase the understanding of the structural condition. The assessment of local pipe coupons or short pipe sections does offer useful information on both internal and external structural condition, however it can be very heavily biased to the local environmental conditions and may not be a true reflection of the overall condition of the main.

A more complete understanding of the condition of a utility's transmission and distribution system is essential for the most efficient prioritization of maintenance and correct decision making processes for rehabilitation and other associated activities. The worst parts of the network could be targeted through a better-informed decision process. There is also the potential to save both money and time either by reducing the amount of pipelines replaced prematurely or needlessly by a better understanding their condition up front, whilst also not allowing pipes to reach a dilapidated state causing future failures and potential widespread damage. For most utility companies, this is probably the strongest driver for using condition assessment techniques.

The use of a comprehensive decision support system (CDSS) would benefit many organizations carrying out water pipe rehabilitation or replacement (Deb et al. 2002b). The system created for AwwaRF by Weston Design Consultants and Virginia Polytechnic Institute and University (Virginia Tech) presents the user with a question and answer interface, and tables where the user can enter data about the proposed work. Once all the data is entered, the program presents the user with several viable options for pipe materials and sizes, installation method and how much each option would cost. The output is also saved as an Excel™ (Microsoft®, Seattle, Wash.) spreadsheet file so it can be opened on any computer with the spreadsheet program installed. This would speed up the decision process. At present the majority of utilities and water companies do not employ such a system and just rely on traditional methods such as using historical data or the experience of their professionals to make these decisions.

A product that has been on the market for some time is the WRc Analysis of Techniques for Effective Rehabilitation Financed Over Whole Life (WATERFOWL). This is a data intensive package designed to assist the operations manager in selecting the most appropriate method for rehabilitating a section of water pipe. It takes into account the original pipe material, any new diameters required, the existing hydraulic capacity, ground conditions and the current depth of the existing pipe amongst many other variables. Once the data has been fed into the software it then prints out a number of potentially suitable rehabilitation techniques for consideration.

A predictive analysis tool has been tested in Melbourne, Australia. It is the Watermain Renewal Analyzer & Prioritisor (WRAP) and it is an Access™ (Microsoft®, Seattle, Wash.) database developed by South East Water's (SEWL) Water Renewal Team in 1997. Data concerning a pipeline is entered into the database where it can be analyzed and given a score depending on the age and material of the pipe, severity of damage, the population and property surrounding the pipe and other factors (Geehman 1999).

A pan-European research effort introduced at the NO-DIG conference in Copenhagen in May 2002 is considering Computer Aided Rehabilitation of Water Networks (CARE-W). This aims to enable the end user to create a database of performance indicators from which they can predict the structural failure and hydraulic performance of pipes in their network and then produce strategies for planning and investment of short and long term rehabilitation schedules. Since there are different regulations and procedures in each region the program uses several models, from which the user can select the most appropriate depending on the amount of data available. The prototype of the CARE-W program will be tested in twelve cities across Europe with different environmental conditions, the findings will be examined in higher education institutions and the project will run until Spring 2004 (Sægrov et al. 2002).

Predicting Future Failures

Condition assessment can also be used to help predict the remaining useful life span of a corroded or cracked pipe. This can be calculated by analyzing the size of the defect and working out how many pressure cycles it will take to cause the defect to become large enough to allow the pipe to fail (Hodges 2001). Also understanding the structural condition at any particular moment in time and then predicting the remaining life can be better quantified by subsequent assessments within a particular time duration. This will give a much better indication of deterioration rate and will help a company's management team to better plan the investment for their pipelines.

A model exists that can be used to prioritize the repair or replacement of cast iron transmission and distribution pipes (Deb et al. 2002a). This model is based on the relationship between remaining wall thickness and residual strength. The environmental stresses exerted on a pipe by external forces (weight of soil, traffic load, and frost load) and internal forces (working pressure and surge pressure) are the variables used to model the stresses exerted on a pipe. These parameters are fed into the model and are compared to the current strength of the pipe and from all this information the Safety Factor (SF) is calculated. If the internal and external forces are similar to the current strength of the pipe ($SF = 1.0$) then a decision needs to be taken whether the pipe should be repaired or replaced.

Assessing Leakage Reduction

For maximum efficiency, it is important that water utilities remain within agreed targets for leakage, or operate at an economic level. Millions of gallons of water are lost every day by the average water utility. The loss of water can cost the water utility millions of U.S. dollars per year in production, cleanup, claims for compensation and loss of consumer confidence. Condition assessment of pipes can also help to verify the severity and the location of leaks so that repairs or replacements can be prioritized and dealt with more quickly and efficiently.

In the U.K., OFWAT has set mandatory targets for leakage reduction for each water company. These targets were introduced as a reaction to the droughts, which occurred in the U.K. during the summer of 1994 where a shortfall in water supply revealed the large amount of water being lost everyday by water companies. After an investigation into the failure to meet the high demand for water, the report submitted to OFWAT highlighted that the main cause of water loss in water pipes was leakage. The benefits of reducing leakage are

- A reduction of water abstraction
- Improved reliability of water supplies
- Less discoloration that occurs when leaks progress to bursts
- Deferral of capital expenditure
- A reduction in operating costs
- Better public profile

Security of Supply and Risk Management

The security of supply in general terms is a measure of the water available to meet current and projected future demand. A useful measure is the ratio of reliable supply to peak demand, which allows prediction of any future shortfalls in supply. Factors to be considered include; population growth, changes in industrial use and contractual improvements, e.g. more connections, improved intermittent supply. Also included in security of supply, and of more immediate concern, is maintaining supply into large zones or for large industrial users or critical customers, such as hospitals.

Domestic, commercial and industrial customers need to have a constant supply of clean water for consumption and industrial processes. The provision of this service can be jeopardized if the associated pipe network is not fit for purpose. The major objective of a water utility in maintaining security of supply must be to determine and understand the level of risk posed to these critical supplies in a systematic and controlled way and to manage the level of risk of loss of supply to these critical customers.

The associated network can be large diameter pipes from water treatment works (Winn 2003, Glozzy 2003), strategic or critical mains (Morrow 2003), or feeds into large distribution zones or to supply large industrial customers. The consequences of failure of these parts of the network can be severe both in terms of interruption to supply to critical customers and the associated problems with the surrounding environment, and disruption to traffic. From a network management point-of-view, there may also be severe difficulties with repairing the damage and maintaining supply to the remaining parts of the network. Condition assessment of these strategic or critical mains would help to reduce the likelihood of such disruption, and allow any remedial measures to be undertaken in a planned and coordinated way.

Asset Valuation

Asset valuation is an important management tool. It assists in the determination and allocation of costs and supports performance/rate of return reporting, resource allocation, shareholder equity and accountability. Asset valuations are required for a variety of reasons, which may include financial reporting, determining appropriate pricing and funding levels, risk management (insurance), asset management decision making and benchmarking. Many infrastructure assets are specialized assets, which are rarely sold on the open market and their potential profitability is not a relevant valuation method. In such cases, their value to the asset owner for the existing use can be best assessed on the basis of optimized replacement cost (ORC) and optimized depreciated replacement cost (ODRC).

ORC is the value of an asset that replicates the existing asset most efficiently, while providing the same level of service. The optimization process takes into account technological change, over-design, redundancy and system reconfiguration. ODRC is the optimized replacement cost after deducting an allowance for either aging, or consumption to reflect the remaining or economic service life of the asset. Generally replacement cost is calculated at asset component level (normally the level at which the asset is replaced). Depreciation is the allocation of the cost of an asset over its useful (economic) life. Depreciation is not charged to provide for replacement of assets but rather to allocate the cost (less any residual value) over the useful life. Depreciation is important to reflect the appropriate carrying value in the fixed asset register and statement of financial position.

In calculating depreciation of an asset, the determination of the useful life of that asset is necessary. In the absence of detailed condition assessment to determine appropriate useful lives for assets, sample listings of useful lives may be used as a guide. However, any more sophisticated approach will utilize condition grades, useful asset lives from industry guidelines (would need to take into account local factors), replacement costs of asset components and key performance indicators to assess useful life. The base life is a key variable that significantly affects the remaining life, as well as the determination of the deterioration rate. The use of condition assessment tools and techniques will allow much more accurate asset valuation by helping to establish the base and deterioration rates for underground assets (which can account for around 70% of the total asset value for a water utility).

Reduction of Operational Expenditure Costs

The cost of maintaining water pressure will increase if there is tuberculation present within a pipe as this reduces the equivalent diameter ($d_e = 4 \times$ Cross-sectional Area/Perimeter) and therefore reduces the volumetric flow rate and hydraulic pressure. Any leaks within the pipe network will also contribute to a loss of pressure. To compensate for the loss of pressure at the tap, a water utility has to increase the pressure at source. This means increasing the power fed to its pumps or buying additional pumps to cope with the increase in pressure demand, running pumps more consistently (perhaps even at peak electricity tariff times) all of which increase pumping costs. Although any excessive levels of tuberculation may be identified through hydraulic gradient and flow patterns, condition assessment techniques will assist with identifying the actual levels of tuberculation in water mains and help to better target any remedial works to reduce the problems.

The operational expenditure involved in repairing a large diameter main failure can also be quite considerable, especially if it is on a strategic main or in a congested area. Condition assessment may allow areas where potential failures could occur to be identified and appropriate remedial action can be taken in a controlled way. This work could be done to suit the utility and local conditions. As a burst can occur at any time, isolating the area and maintaining water quality and supply by whatever means are the first priorities. Then repairing or replacing the damaged section as quickly as possible is next, often with a large number of crews and equipment at premium or overtime rates. Therefore any way to make this process more efficient and less costly would be a direct benefit to the utility. However, the cost and disruption of performing the survey would need to be assessed against the cost, disruption and service failure that a burst may produce.

BUSINESS JUSTIFICATION

Deferral of Capital Expenditure

Most of a water company's fixed assets are underground pipes, some of which have been there for more than a century. A water company's records of these assets will be less reliable the earlier the pipes were first installed. The records are updated only if one of these older pipes fail in some way and require remedial work. Condition assessment can not only assess the state of the pipe infrastructure, but if it were equipped with a locating device, it could also reveal the exact position of a water company's pipes.

Some studies recommend that there should be more research into corrosion measurement (Kirmeyer, Richards, and Smith 1994). The water industry should express its interest and invest in improving and adapting non-destructive testing techniques for use in pipeline condition assessment, particularly for tools that could comprehensively evaluate pipe condition (Jackson, Pitt, and Skabo 1992). Some developments in non-destructive techniques for water mains has since taken place, and these are included in Chapter 3.

Resources should be put into in-line periodic condition assessment programs so that the lifetime of pipes can be estimated. If water companies had programs of this nature in place they would be able to assess the state of their pipes and create a program of repair, rehabilitation and replacement for their entire network. They would be able to identify and prioritize which pipes needed immediate attention and those that did not, and so would be able to allocate resources accordingly (Rajani et al. 2000). They would also be able to judge the risk posed by dilapidated pipes and rank them on the severity of damage based on the data gathered by the assessment tool. This would also be factored into the repair schedule, enabling a water company to shift from a reactive to a proactive operating repair system and repairing only the damaged sections of pipe rather than going to all the cost and effort of replacing an entire length of pipeline. Some water companies have already saved a great deal of money by operating in this way (Mergelas and Kong 2001).

Traditionally when a water company surveys a section of its pipe network, it may take a sample by extracting a pipe coupon or removing a short pipe section for analysis in a laboratory. An entire section of pipeline may then be replaced on the basis of the results of the analysis of the short pipe length. The rest of the section could still be in good condition and still have decades of service left before it requires replacing however; it could still be replaced prematurely at great expense to the water utility.

Condition assessment of the entire section would identify whether lengths of pipe actually need to be replaced now, or whether localized sections could just be repaired to prevent future failures. This would mean capital would not be spent prematurely or unnecessarily and could potentially save the water utility millions of U.S. dollars per year.

The potential consequences of failures within large diameter pipes is illustrated by the 24-inch (600-mm) steel pipe situated on Lord McGowan's Bridge near Middlesborough, Cleveland, U.K. (Appendix H, Case Study 1). It is also located adjacent to a large industrial estate comprising of ICI and DuPont chemical works amongst others and a railway line. The cost of replacing the pipe was $475,000 (£300,000) and the cost of compensating domestic and industrial customers and Railtrack was an estimated $1,275,000 (£800,000) (Morrow 2003).

Reduction of Capital Expenditure

By only targeting the areas in need of replacement instead of replacing entire lengths of pipe, a water utility can significantly reduce its capital expenditure every year. Condition assessment can help with this spending reduction by assisting with the identification of the pipe sections that actually need to be replaced. This saves time, manpower and money. This would particularly apply to trunk mains as these are the most expensive to replace.

The average cost to replace a 12-inch (300-mm) water main is $57.12/ft (£35.44/ft) in the U.K. Relining a 12-inch water main costs just $32.78/ft (£20.34/ft) at 1997/98 prices, (Sutton and East Surrey Water plc. 2000). Condition assessment can assist with making the correct renew or rehabilitation decision with the associated cost benefits.

Cathodic protection is another cost effective solution to pipe maintenance. For example the cost of replacing 0.625 miles (1 km) of 12-inch water main in Ottawa would be $780,878, but to retro-fit a pipe with a sacrificial anode system three times would cost $79,862 in total and would give forty to fifty years of protection. This represents an 89% saving over replacing the pipeline (Raymond 2002). The state of the sacrificial anode could be monitored to inform the water company when it needs to be replaced.

Reduction of Operating Expenditure

Leakage Reduction and Associated Savings

Leaks are responsible for losing millions of gallons of water per day. For the average U.K water utility, 14% of piped water is lost and in some areas of the world up to as much as 65% of water production is lost every day. If an oil company lost that much oil per day it would soon be out of business and subject to legal action, heavy fines and its directors could possibly face a custodial sentence. Water companies cannot afford to have that much volume of unaccounted water, but many do whilst others are approaching their economic levels of leakage. Leaks are caused by corrosion, pin holes, cracks, mains bursts, faulty joints and pressure surges amongst other things. Many small leaks can continue undetected for considerable periods of time and so the value of overall lost volume of water can become considerable. This is illustrated in Figure 2.2 for various leak sizes.

The cost of lost water could be used to justify the investment required to perform the condition assessment and the subsequent repair. A basic model has been produced to allow the costs of leakage screening (using arial imagery) and subsequent field investigation and repair to be plotted against number of leaks detected. The model output is a break-even chart indicating the number of leaks that need to be detected and repaired for payback of screening and repair costs in one year. Any further leaks detected and repaired would result in a direct reduction in operating costs. This is shown in Figure 2.3. The assessment also has the advantage of accurately pinpointing the leakage, and so minimizing excavation.

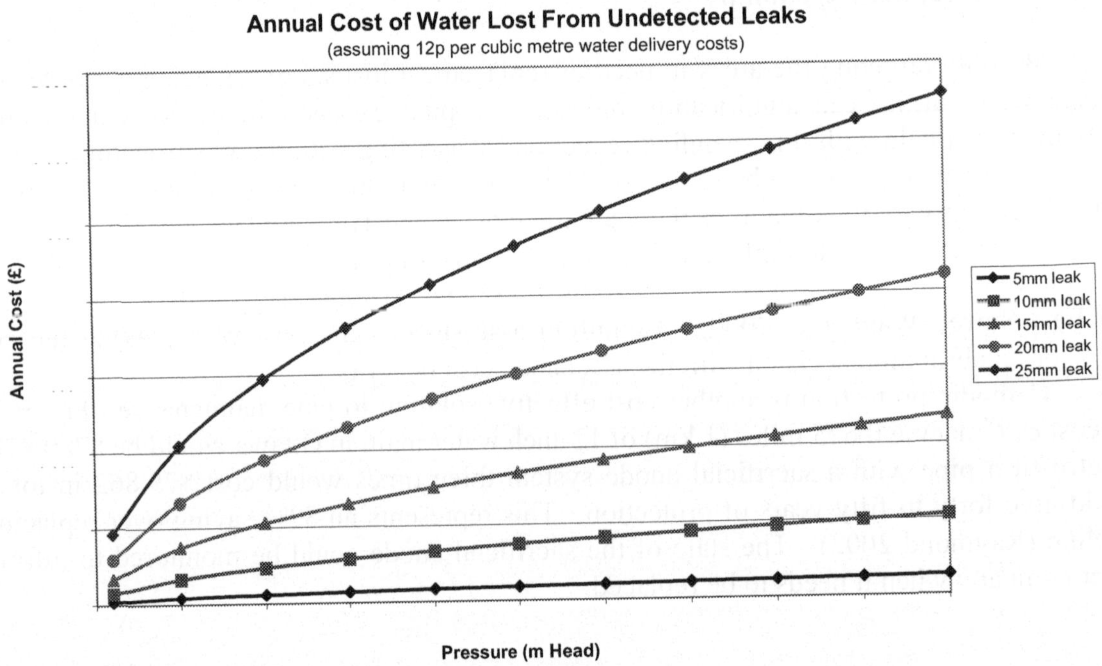

Figure 2.2 Lost water cost for various operating pressures (*Source*: ISSquared Ltd., 2002)

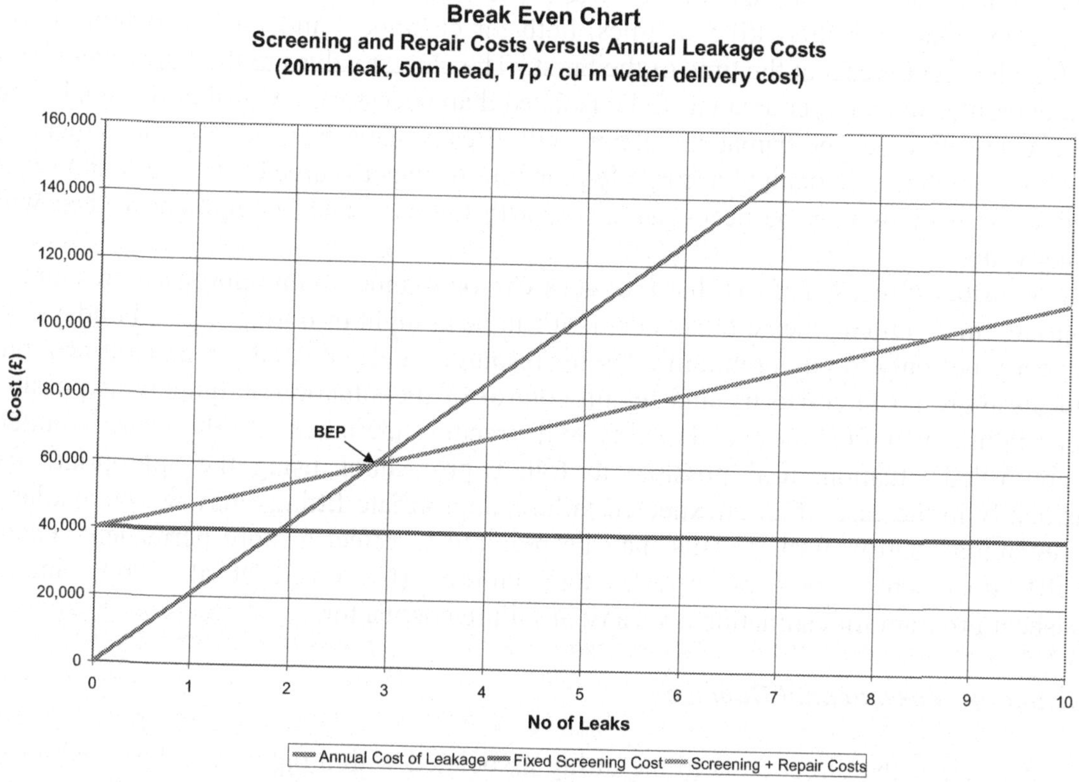

Figure 2.3 Break even chart (*Source*: ISSquared Ltd., 2002)

Further savings could be made in areas where the resource is quite scarce. If reducing leakage could save more water, it may be that further sources do not need to be exploited to meet current or future demand as the network expands. This would offer considerable capital expenditure savings if new facilities were not required.

Benefits of Proactive Repairs

Unless a particular area of the network has a history of pipeline failures, it is difficult to predict where future failures are going to occur. This is because there are a number of factors that are essentially unknown such as pipe condition, composition of surrounding ground, and future ground movement. As such, at the moment, most water utilities react to pipeline failures when they occur, and try to repair the damage as quickly and efficiently as possible. The costs of trying to understand where failures may occur throughout the whole network would be excessive and uneconomic, but for selected areas of the network (for example critical or strategic trunk mains with high consequences of failure), there are a number of benefits which may be considered.

The primary benefit is to reduce the number of unplanned outages with an associated reduction in overall disruption to the supply, the surrounding area and to customers.

A secondary benefit is that proactive repairs would allow better control of the situation. It would not be necessary to take precautions over the possible effects on water quality through disturbances within the main (leading to brown water problems) and through infiltration of possible contaminants at the site of the break. It would also make it possible to plan any provision necessary for alternative supplies, both for customers and for fire fighting purposes, rather than having to react at the time of the burst. The closest valves to the failure may be found to be inoperable, and a larger area has to be isolated than necessary. A further potential benefit is reduced excavation, as the damaged section will already have been identified. When a burst occurs it is sometimes difficult to accurately pinpoint the exact source of the break and several or extensive excavations may be necessary along with the associated equipment to deal with the escaping water.

A further direct benefit is that the work can be planned with appropriate resource at an appropriate time. Often when a break occurs it can be at night or during public holidays when it is necessary not only to pay premium rates for resources and potentially repair fittings, but also to allocate more resources to try and regain control of the situation as quickly as possible. A further potential benefit is the actual quality of the repair undertaken. As the repair is undertaken in a controlled situation, it is possible to follow procedures using the appropriate fittings. Occasionally in the case of an unexpected failure, appropriate fittings may not be available and after regaining control, further visits may be required to effect a more permanent repair. The AwwaRF document "Costs of infrastructure failure" (Cromwell et al, 2002), includes a spreadsheet program for estimating a variety of failure cost factors.

Reduction of Consequential Damage

As well as the direct costs to the utility associated with a pipeline failure such as cost of lost water, excavation, manpower, mobilization of equipment, repair fittings and reinstatement, there are also a number of other potential direct costs to be borne by the utility. In the immediate aftermath of the failure, there can be a large cost associated with the subsequent clean-up operation in the surrounding area, and in some cases this may include the inside of any properties which have been flooded. There may be debris from loosened soil and aggregate contained within the water that may have entered properties and been deposited onto streets. There can be odor from the dirty water, which penetrates fabrics such as carpets, upholstery and clothing and therefore there may also be the need to remove and replace any damaged soft furnishings from these properties. Although these costs may be covered by insurance, there may be an effect due to increase in premiums for claims and also a direct cost through utility staff having to arrange and co-ordinate these activities.

Other direct costs to the utility will include any site investigations required to try and determine the cause(s) of the failure, and whether any further remedial measures are required to stabilize the area and ensure that further failures do not occur. The incident will need to be documented, and if necessary reported to the regulator, at a further cost to the utility. There may also be other costs associated with the legal aspects of the incident and trying to determine whether any third parties were responsible (either utilities or construction companies) or liable.

Following the direct costs to the utility, there are a number of potential indirect or social costs, which may also need to be included. At the time of the incident, there may be considerable disruption caused not only to customers, but also to local businesses and traffic. It is difficult to put an actual value onto this disruption, but if local businesses are unable to function because of the failure, they may make a subsequent claim against the utility.

The environmental impact of leakage or mains pipe burst can be substantial, particularly within urban areas. Water can erode the surrounding soil and aggregates (Insituform® 2002) and consequently weaken any adjacent buildings and structures. Increasing the water content in soils causes the soil particles to slip against each other and reduce the total soil volume. This can lead to subsidence and therefore damage to pavement structures and buildings (Tawfiq 2002). There may also be claims submitted for damage by other utilities if their plant or supply is effected by the pipe failure, and these costs can be quite significant. There may also be subsequent claims for damage to properties, or weakening of structures which may lead to further damage. There may be costs for damage to local flora and charges associated with the discharge of the treated water.

The risk of occurrence of consequential damage needs to be assessed, but if all of these direct and indirect costs could be quantified as actual costs to the utility in terms of time and money, in the case of large diameter pipe failures, they may prove to be quite considerable. In an attempt to improve the prioritization of asset renewal on large diameter mains, criticality models have been developed (PPK Consultants Pty. Ltd.) which may also be applicable to assist with quantifying potential financial consequences of pipe failure. Negative publicity associated with these incidents could also have a longer reaching effect on the companies' reputation. The majority of these consequential costs could be avoided if proactive repairs are undertaken following condition assessment activities.

Economics of Condition Assessment

The value of condition assessment can be calculated in terms of cash. There are a few economic models that can be utilized to do this.

Net Present Value

One of the most common is the Net Present Value (NPV) standard formulation. This is the value of an investment, in this case performing condition assessment; it uses a formula based on the initial outlay, the periodic return on that outlay and the depreciation of value over time. The formula is as follows:

$$NPV = \sum_{i=1}^{n} \frac{v_i}{(1+r)^i}$$

where NPV = Net Present Value
v = Value
r = Discount rate

The first value used in this formula would be the cost of performing the condition assessment; the other values would be the periodic savings gained through conducting the condition assessment (e.g. capital expenditure deferral and reduction, operational expenditure reduction, knowledge value). These values could also be costs (e.g. repairs or replacements in order to rehabilitate the pipe). The discount rate is a result of economic factors such as the rate of inflation.

For example if condition assessment was carried out on a 1 km section of pipe. The inspection and rehabilitation costs $10,000, and say the income on this section of pipe was $5,000 per year and the inflation rate remains at 2.5%, the NPV of the condition assessment

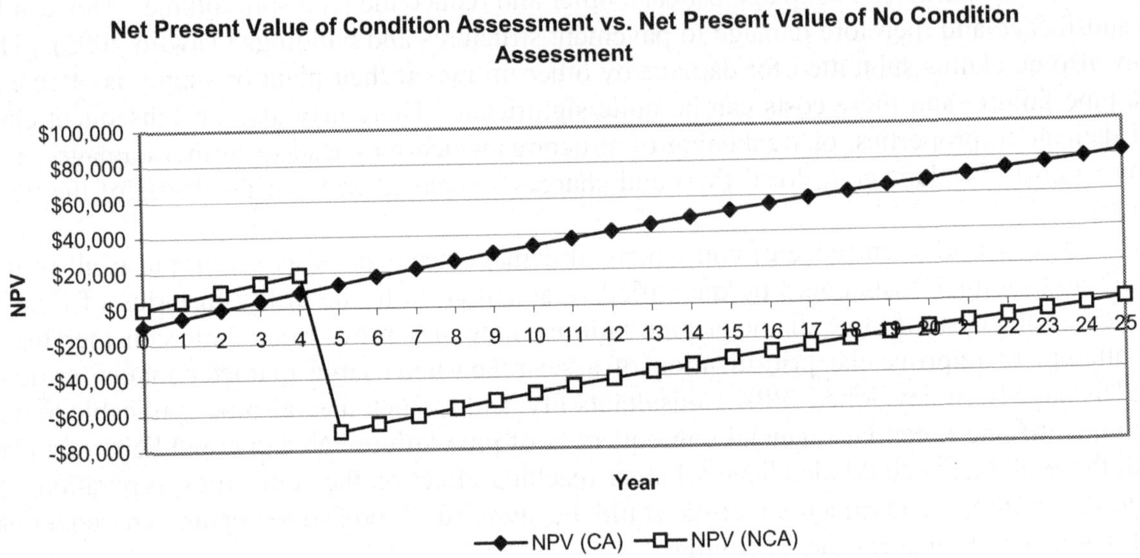

Figure 2.4 Net Present Value comparison

would be as shown in Figure 2.4 (CA). However if no condition assessment was carried out and the pipe failed and caused $100,000 worth of damage in the fifth year then the NPV would be as shown in Figure 2.4 (NCA). However, the risk of occurrence of an actual pipe failure also needs to be considered, as the pipe may not fail within the 25-year period.

As can be seen from Figure 2.4, calculating the NPV over a twenty-five year period shows that with condition assessment, although it will cost a utility initially, the cost of the condition assessment is recouped within three years and continues to generate revenue over the next twenty-two years. If no condition assessment is carried out and the pipe fails after five years and costs $100,000 to restore normal service and clean up then it will take more than twenty years to recover the cost of that failure from the revenue generated by the pipe.

Inspection Value Method

The inspection value method (IVM) defines the value of inspection as the benefits minus the costs of inspection. The benefit of inspection is defined as the product (ΔP (i.e. the reduction of the probability of failure) × the consequences of failure). In order to define the benefit attained from the data acquired from a particular inspection approach, the measurable performance characteristics of the device (e.g. speed, precision, sensitivity, accuracy, probability of detection and coverage (i.e. spatial, temporal and failure mode)). These must be translated into values of ΔP, if they are to be useful for estimating the benefit of inspection. It is also necessary to estimate the consequences of failure (CF) in order to calculate the benefits, but CF does not depend upon the performance of the inspection method. The IVM consists of equations for the benefits and costs of inspection.

Determining whether the benefits are less, equal or greater than the cost of the inspection, enables a justification of the decision for or against the inspection to be made. Factors directly affected by the inspection method that are addressed by the IVM include the efficiency of detection, which is the product (probability of detection of critical flaws (POD) × coverage) and the costs of inspection (Wall and Lilley 2002).

BUSINESS REQUIREMENTS

Introduction

This section considers some of the requirements of the water industry for any potential condition assessment tool. A discussion of water mains general population including pipe materials, diameters and failure rates is provided followed by consideration of the different failure modes for the various pipe materials and what may be the desirable features of any condition assessment tool. Consideration is also given to the potential generic strategies that water utilities may adopt for condition assessment activities.

The average cost of replacing the water pipe network to comply with the required standards in the U.S. over the next twenty years has been estimated to cost almost one trillion dollars. The potable transmission water pipe network would require around $325 billion to upgrade and replace failed pipes, which equates to $16.25 billion annually. There is currently an $11 billion annual shortfall between current spending, and the overall necessary spending (Neukrug 2001). Some of these costs could be offset by undertaking condition assessment and either rehabilitating or relining damaged pipes, which would otherwise have been replaced.

General Market

The average water utility has 1,628 miles (2,620 km) of distribution piping, and produces 95 million gallons of water per day in North America. In Europe, the average water utility has 3,067 miles (4,935 km) of distribution piping, and produces 80.8 million gallons of water per day (Deb et al. 2002a). Most are made of traditional materials such as cast iron (48%), ductile iron (19.2%), and asbestos cement (AC, 15.1%) (Kirmeyer, Richards, and Smith 1994). New pipes are being laid every year, with some made from modern materials for example prestressed concrete, medium-density polyethylene (MDPE), high-density polyethylene (HDPE) and non-plasticized polyvinyl chloride (uPVC). One manufacturer makes HDPE pipe from 3 inches (75 mm) up to 63 inches (1,600 mm) in diameter. So plastic pipes could be suitable for both transmission and distribution mains. What is important is to standardize on the new materials now being installed as in the future, considerable stocks of repair fittings may have to be carried by water utilities to accommodate the various sizes and types of materials now in use.

On average, North America suffers 22 failures/100 miles/year (13.7 failures/100 km/year), whereas Europe suffers 80.5 failures/100 miles/year (50 failures/100 km/year) (Deb et al. 2002a). This works out on average each utility suffers 358.16 breaks/year in North America and 2,468.5 breaks/year in Europe.

The Water Supplies Department of Hong Kong suffered 2,236 mains breaks in an eight-month period in 1995. The total repair costs were around $9,058,600. Of these there were 105 serious water mains bursts each of which cost over $21,800 to restore supply, and in total these bursts cost $3,461,700 to completely rectify. These major pipe bursts accounted for only 4.65% of incidents, yet they accounted for 41.44% of the total costs (Legislative Council of Hong Kong 1995).

Failure Modes of Pipelines

Correct material selection, design and appropriate bedding and backfill compaction should allow a reduction in the numbers of pipeline failures in the future. However, pipeline failures are generally attributed to a number of contributing factors over a period of time rather than a single immediate event. They can be due to inappropriate pipe design and unexpected variations in the operating environment or a combination of these. Incorrect storage, handling, transportation and installation of the pipe can lead to consequential failure due to damage initiated either at the pipe surface or at the protective coating. After commissioning, the pipe is subjected to a continuous degradation process. Although the initial design should have allowed for this under normal operating conditions, internal and external corrosion can occur in metallic pipelines and ground contamination or water quality can affect the integrity of mineral and plastic based materials. Coupled with changes in operating pressure cycles or ground condition during the lifetime of the pipe, extra stresses can be imposed on the pipe that may then lead to premature failure.

The purpose of any condition assessment tool should therefore be to establish the level of deterioration a pipeline has suffered and allow a prediction of its future useful life. To be able to do this, the tool will need to be able to detect the major potential failure modes for the pipeline materials in use today.

For the cast iron materials, internal and external corrosion is the most common contributors to pipe failure, both through general corrosion and pitting corrosion. For small diameter pipes, external-bending loads can also have an influence when a pipe fails and for larger diameters, crushing can lead to longitudinal cracks and ultimately to failure. Due to the pressure containment capabilities of these materials, internal pressure generally has little influence unless the other factors are involved. Steel can suffer from similar problems, and the geometry of the pipe can vary depending upon local ground loading. Any condition assessment tools would ideally need to be able to determine the remaining wall thickness and whether corrosion is internal or external (a possible indication of lining or coating breakdown for more modern materials). Tuberculation measurement is also a good indication that pipe deterioration is underway as it highlights areas where corrosion may be taking place.

For mineral based materials, such as pre-stressed concrete and asbestos cement, softening of the pipe structure can occur through lime leaching when in contact with certain soil and water types. For pre-stressed concrete cylinder pipes (PCCP), this process also involves corrosion of the wire reinforcement with a corresponding loss in structural integrity as the wires fail. Some support to these pipes is offered by the pipe bedding and it would be important to establish any voids or uncompacted areas.

For plastic materials, crack initiation through surface damage can lead to premature pipe failure depending upon operating pressures and ground loading. Chemicals in contaminated ground can also lead to a weakening of the pipe structure, and again it is important to understand the level of support provided by the backfill.

Utility Strategies

During the workshop, a number of pipe condition assessment strategies were identified. For one utility, the strategy focussed upon transmission mains larger than 24 inches in diameter, due to the deteriorating condition of riveted and welded steel lines from corrosion (Gomez

2003). The condition assessment process involves combining information generated on a number of parameters. The pipeline history incorporates data on pipe age and material, previous inspections and leak reports and condition of valves and the location of the pipeline includes depth, water table, traffic loading and potential land slip data. Lines are also pressure tested against valves to 1.5 times their working pressure and any leaks are located and repaired. The company is also generating soil corrosivity maps.

For another utility, the replacement and rehabilitation program covers a range of pipe sizes and around 1% of the total network length was included in 2002 (Bush 2003). The assessment process includes gathering information from several sources. Maintenance history is considered, and trend analysis performed at regular intervals. Customer water quality complaints can highlight areas for clean and line projects and pressure complaints can lead to capacity upgrades. Installation information is considered covering manufacturing practices and any changes in specification and construction practices. Hydraulic tests can be performed to assess flow capacities and to indicate interior condition and tuberculation. CCTV can be used to provide a record of interior condition, and pipe coupons can be taken to assess condition and mechanical characteristics. Infrared thermal imaging has also been utilized on a trial basis to locate leaks. Only the higher value transmission mains require multiple assessments.

In the development of any condition assessment strategy, the utility needs to understand what data is required to manage the network efficiently and what can be used to assist in meeting the agreed levels of service. This may vary for each utility, depending upon operating conditions, regulation, extent of network, public or private ownership etc., but the fundamental data required for managing a water distribution network is generally the same. Consideration also needs to be given to the value that condition assessment activities can bring to the business. Even if the majority of failures are found in small diameter cast iron pipes, it may be that larger diameter ductile iron pipes are more critical to the business in terms of levels and consequences of failure (Legislative Council of Hong Kong 1995). An understanding of the priority of needs to the business can assist in targeting any condition assessment activities for maximum benefit. These needs may vary between water quality considerations, pipeline hydraulics, general or detailed pipe condition and leakage evaluation. An understanding is then required of what can be done with the data that is generated, can it be integrated into the other systems already operated by the utility. What useful information and decisions can be made from the data that has been generated, and how can it be stored and updated?

The potential sources of data for use in an assessment of pipeline condition can be varied, but initially, historic data on previous failures, indications of where previous jobs have been undertaken and information from field crews can form the basis of identifying areas to target for further investigation. It may be worth undertaking some initial screening of the area of the network under study. This could take the form of an initial study of a wide area to try and minimize the area requiring further detailed evaluation. Techniques such as thermal imaging have already been used by some utilities to try and identify areas, which are suffering from leakage (Bush 2003). These techniques offer the opportunity to investigate large areas of the network at a relatively low cost per ft compared with the potential costs of the lost water (Reed, Robinson and Smart 2003). Following desk studies of historic data and any initial screening, a more comprehensive inspection can then be undertaken on-site, using the most appropriate technique or device.

A further consideration is the extent of pipeline to be covered by the assessment and the level of data required. If a large area is to be covered, it may be preferential to obtain more limited data to generate an overall view of the network condition rather than a more detailed and hence more costly analysis of shorter sections.

Following the condition assessment, other factors can be taken into account including an assessment of the backfill and an assessment of the loads on the pipeline both internal and external to generate a full structural assessment of the pipeline if this is required. This approach may be beneficial in the case of critical or strategic mains (crossings, single feed, large customer etc.) where the costs can be readily justified (Morrow 2003). Following these investigations, an assessment of the risk and consequences of failure will need to be undertaken (Rajani 2003, Conroy 2003, Wayman 2003).

Consideration can then be given to further condition assessments (on a periodic basis), to determine any changes in the structural condition of the pipeline over time, and to identify any potential problems in advance. In some areas (e.g. where land-slip or flooding may occur), it may be worth monitoring the pipeline either continuously or on a periodic basis, again to identify any potential problems. This can take the form of remote or localized monitoring, and be applied either internal or external to the pipe (Reed, Robinson, and Smart 2003).

TELEPHONE CONSULTATION

The purpose of the telephone consultation was to gain a flavor of the opinions of water utilities to condition assessment and to gauge what current practices water utilities use to assess the condition of their pipe networks in preparation for the workshop held in New Jersey.

The Market for Condition Assessment

This section reviews the results obtained from the telephone consultation of the forty-one water utilities, thirty-seven from North America and four from the United Kingdom. The participants held senior positions in either the operations or engineering departments of water utilities. Where information is available, comparisons are made between North America and the United Kingdom.

The Physical Make-up of Transmission Mains

The majority of transmission mains are ferrous. These ferrous pipes are mainly cast iron followed by ductile iron and steel respectively. The rest are made up of other materials including PVC, PCCP and asbestos cement (AC). The main problems of the pipelines in the water industry are corrosion, tuberculation, breaks and aging pipes. These are illustrated in Figures 2.5 and 2.6.

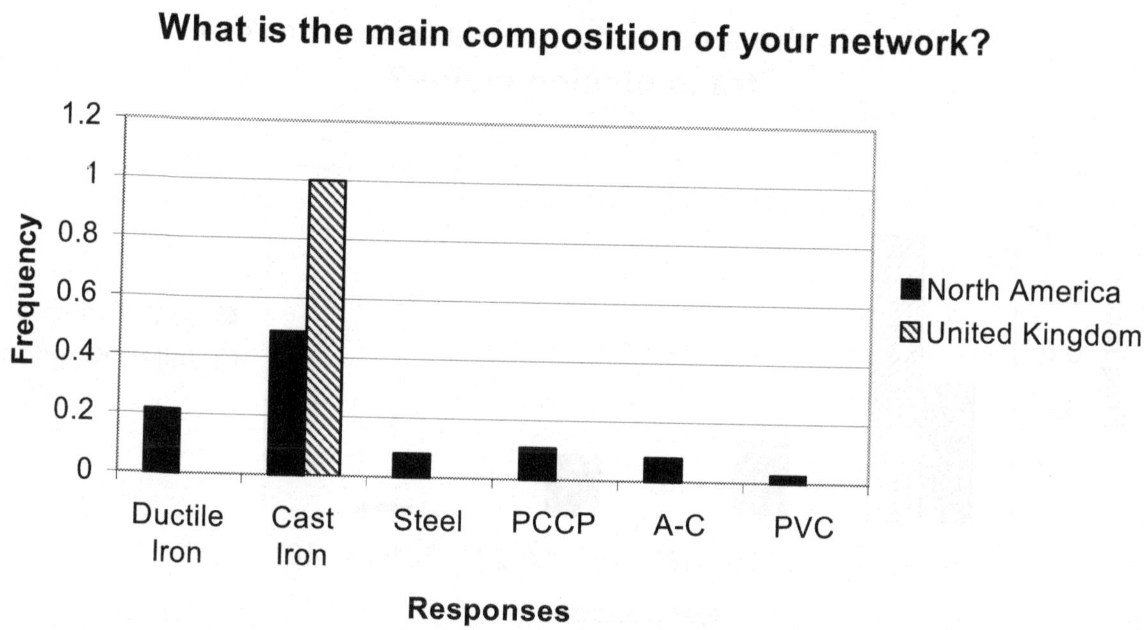

Figure 2.5 Composition of pipe networks

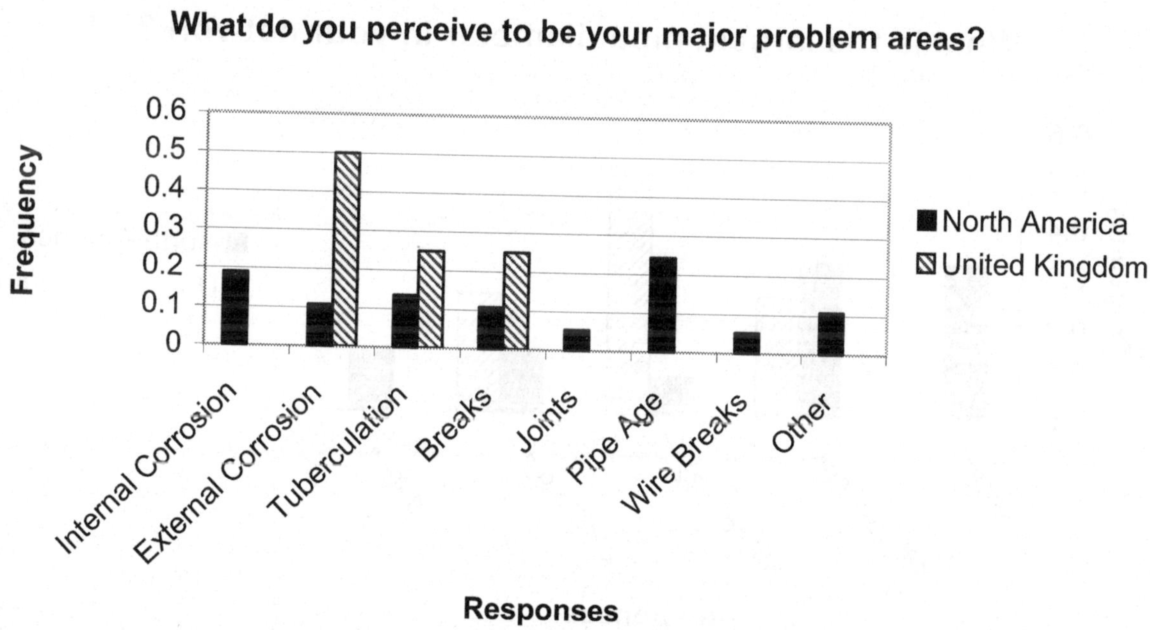

Figure 2.6 Major problem areas of pipe networks

Range of Pipe Diameters for Transmission Mains

The range of pipe diameters for transmission mains was quite large, from 12 inches (400 mm) all the way up to 30 ft (9.1 m), as reflected in Figures 2.7 and 2.8.

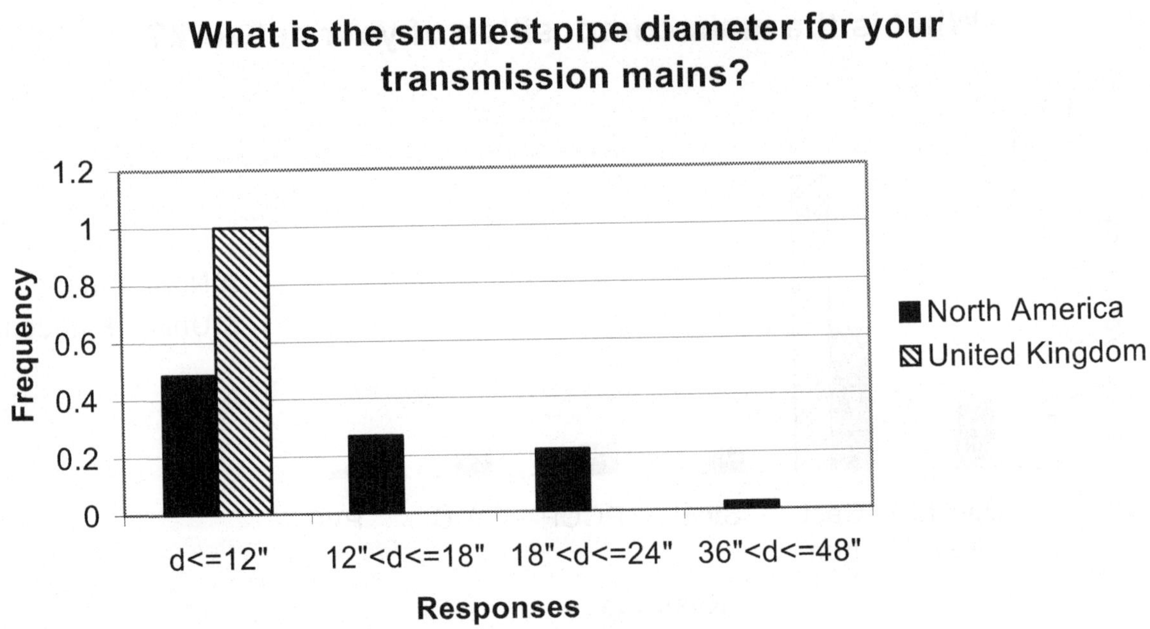

Figure 2.7 Smallest transmission main diameters

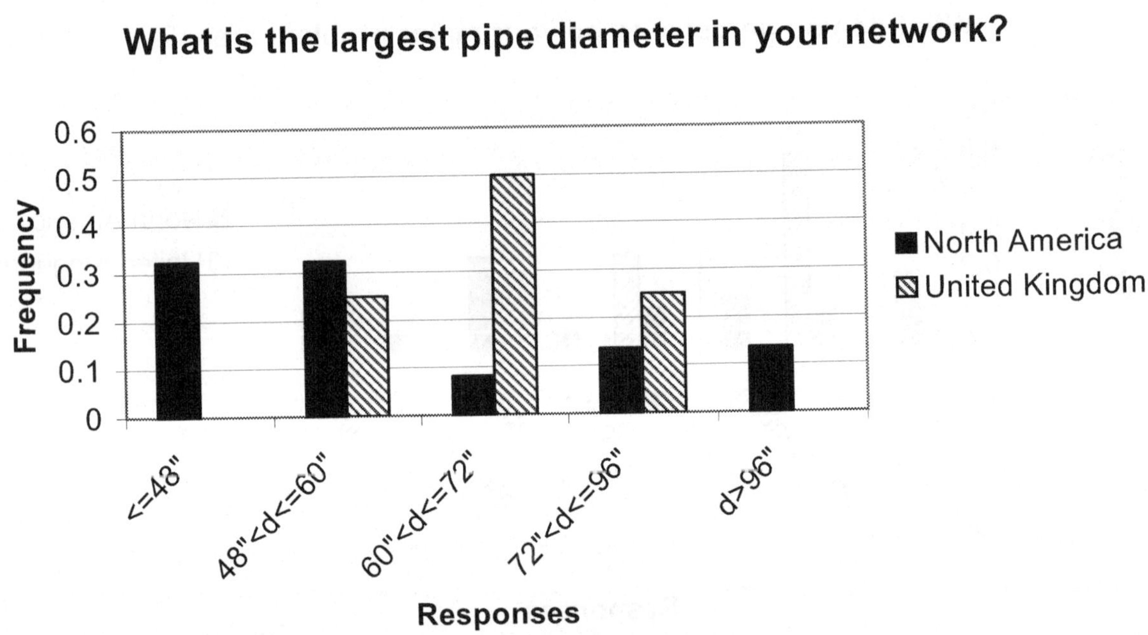

Figure 2.8 Largest pipe diameters

Appeal of condition assessment within the Water Industry

There is considerable interest within the North American water industry for condition assessment. The majority of those who took part in the telephone consultation were keen to know more about the condition of their transmission mains and to learn more by attending the workshop (Figures 2.9 through 2.12).

There were a variety of methods currently used to assess the condition of water pipes ranging from pipe age to complex algorithms; surprisingly two of the people asked did not assess their pipes at all. In addition most water companies would like to know more to quote one person who said "I'm not of the school that says what I don't know, won't hurt me", although three of the respondents did not want to know their pipe network better.

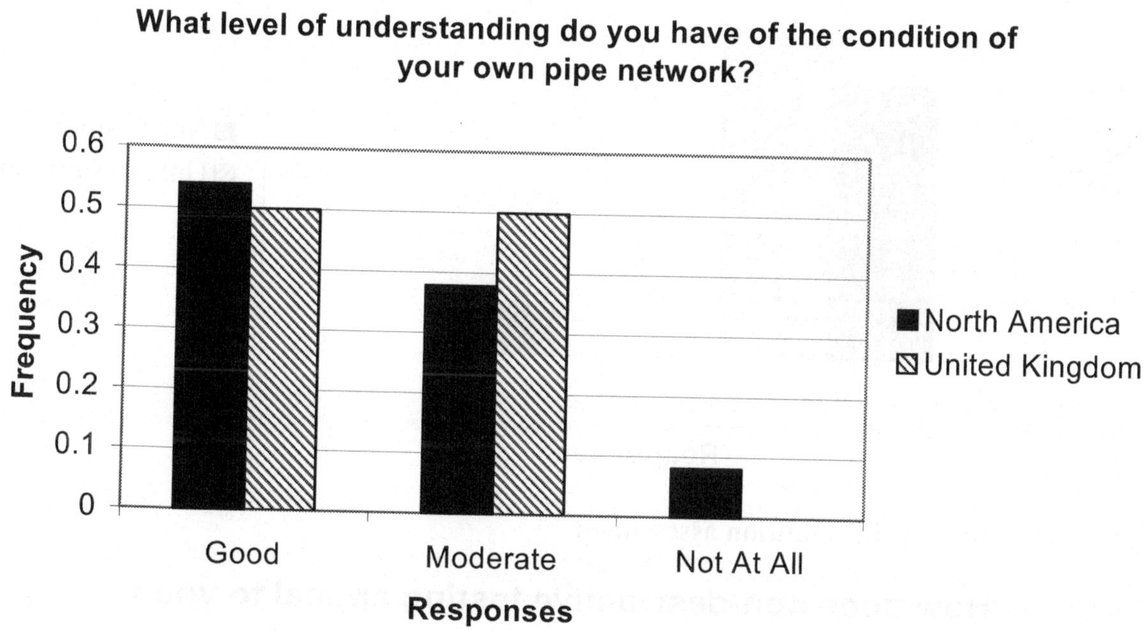

Figure 2.9 Levels of understanding of the condition of pipe networks

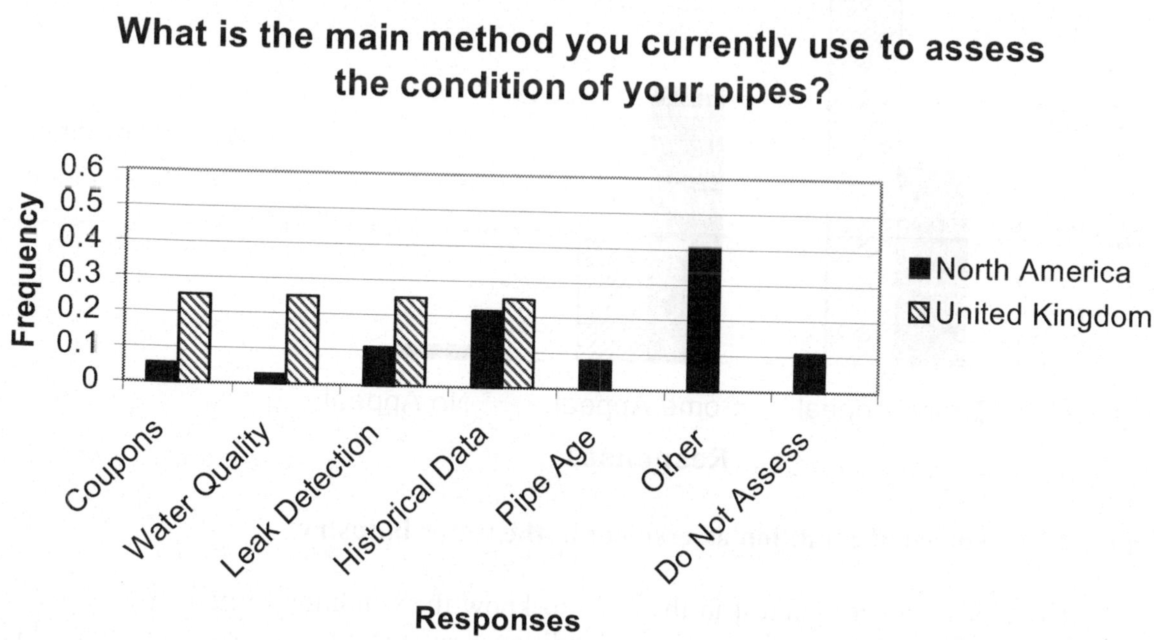

Figure 2.10 Current practices of condition assessment

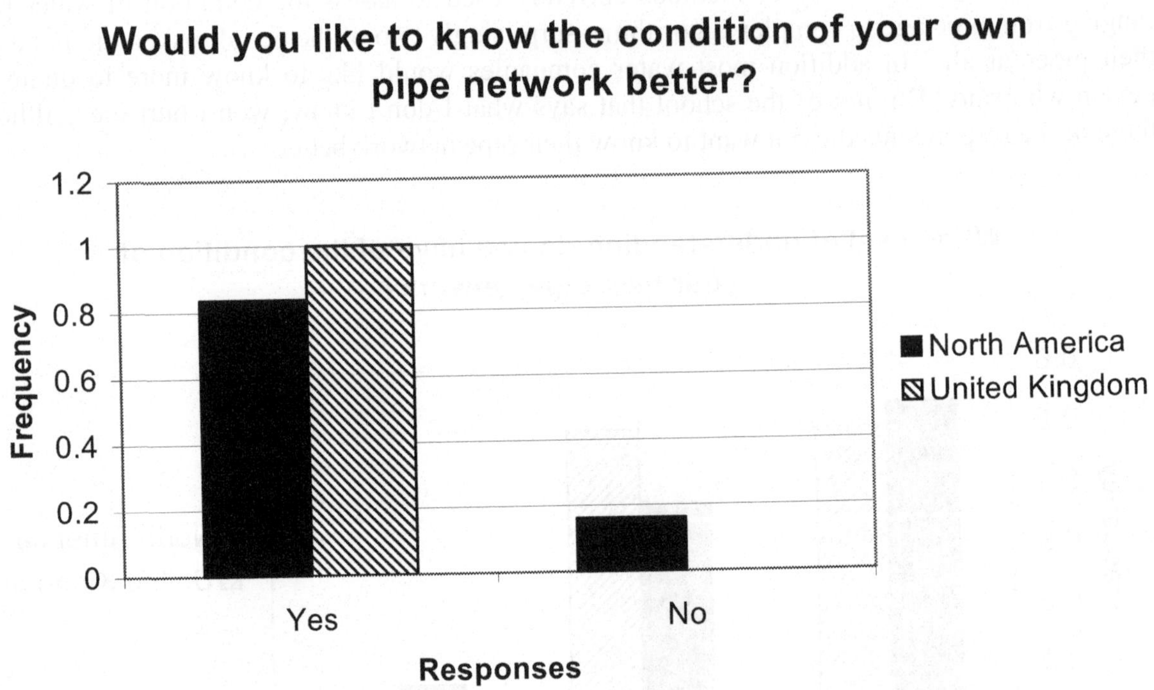

Figure 2.11 Interest in condition assessment

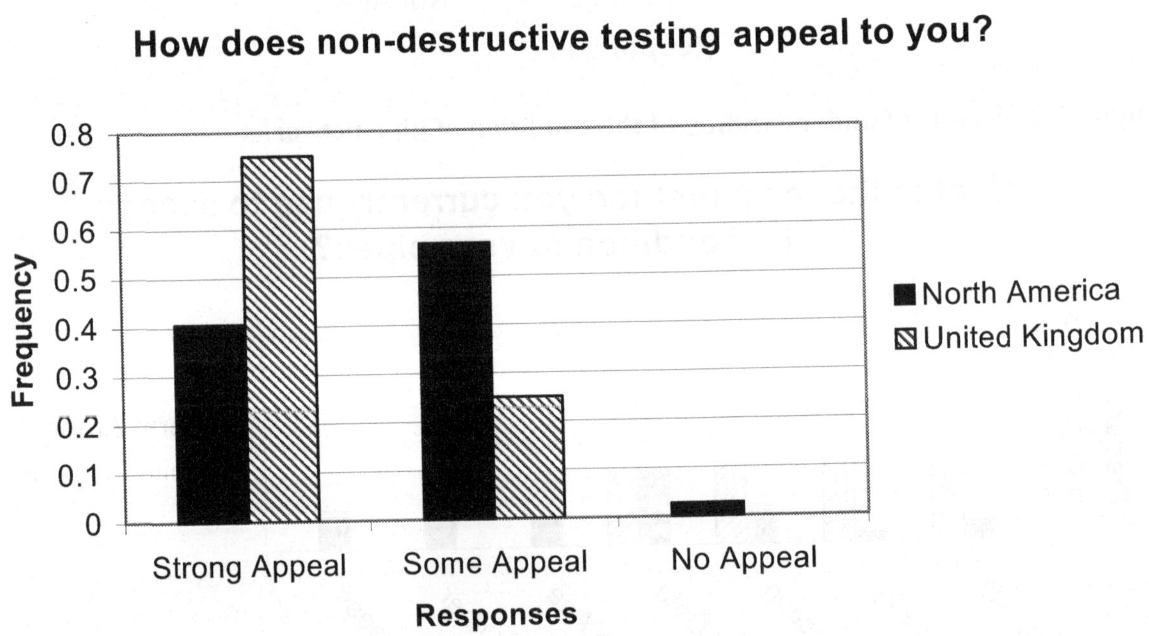

Figure 2.12 Appeal of condition assessment in the water industry

There is a stronger interest in the U.K. to know the condition of networks better and a stronger appeal for non-destructive testing. Condition assessment has at least some appeal to the water industry in North America; most participants felt that they needed to know more about the subject before they could develop a keen interest about NDT (Figures 2.13, 2.14, and 2.15).

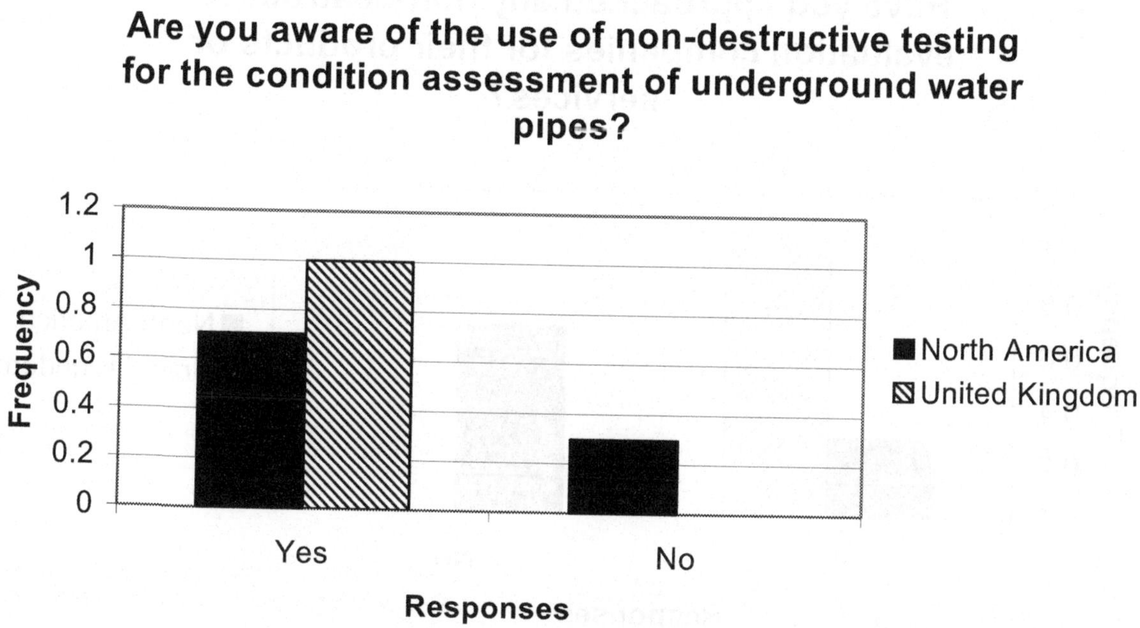

Figure 2.13 Awareness of NDT in the condition assessment of water pipes

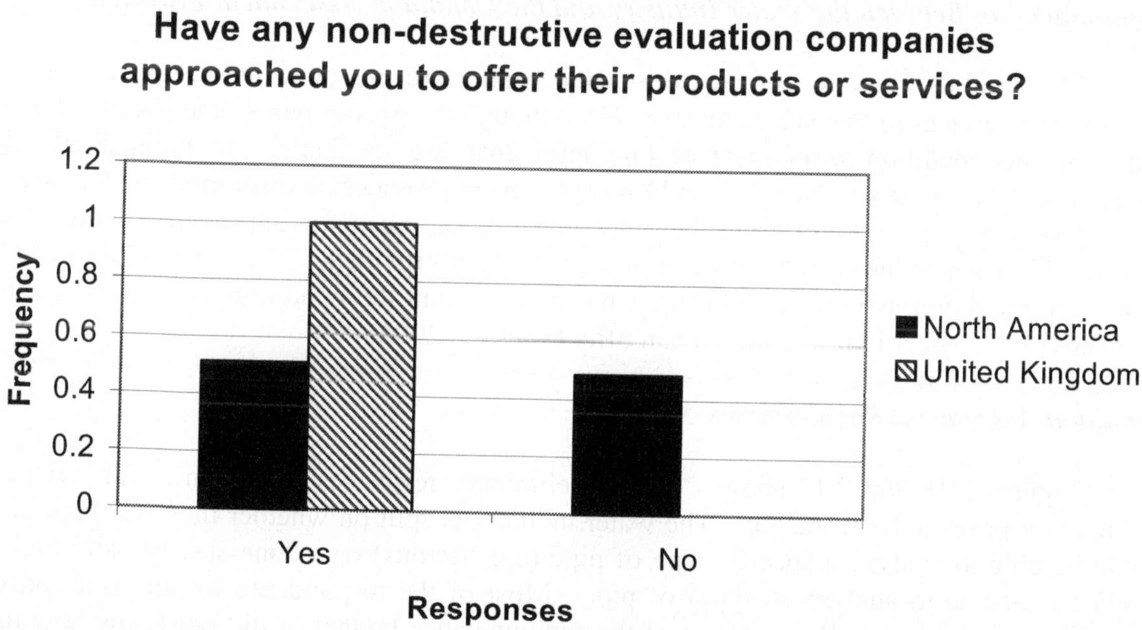

Figure 2.14 Activity of condition assessment companies in the water industry

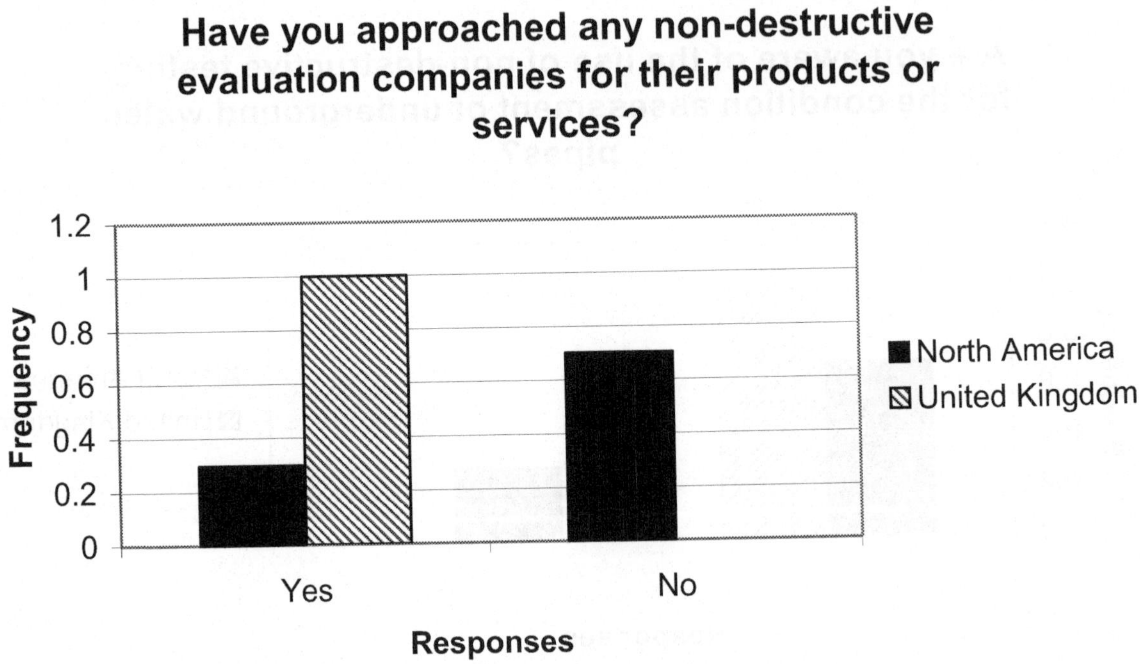

Figure 2.15 Activity of the water industry approaching the condition assessment providers

Communication Between the Water Industry and the Condition Assessment Providers

The condition assessment providers have been more proactive in seeking to sell their products and services to the water industry. Even though the greater part of the water industry is aware of the condition assessment technologies that are available, the number of water companies that have approached the condition assessment providers is disproportionately low.

One key point raised during the workshop was the need for improved communication between the water industry and the condition assessment providers. The water industry has to inform the condition assessment industry what it wants and the condition assessment industry has to advise the water industry what it can offer (Dickson 2003).

Condition Assessment Requirements

Figures 2.16 and 2.17 show a set of preliminary responses taken during the telephone consultation prior to the workshop. The water industry is split on whether they want a tool that would be able to analyze a specific type of pipe (e.g. ferrous) or a "one-size-fits-all" tool that would be capable to analyse all types of pipe. Most of the respondents would either only be prepared to allocate a small proportion of their maintenance budget or did not know how much of their budget to assign to condition assessment. Only one respondent would be prepared to allocate a significant proportion of their budget to condition assessment.

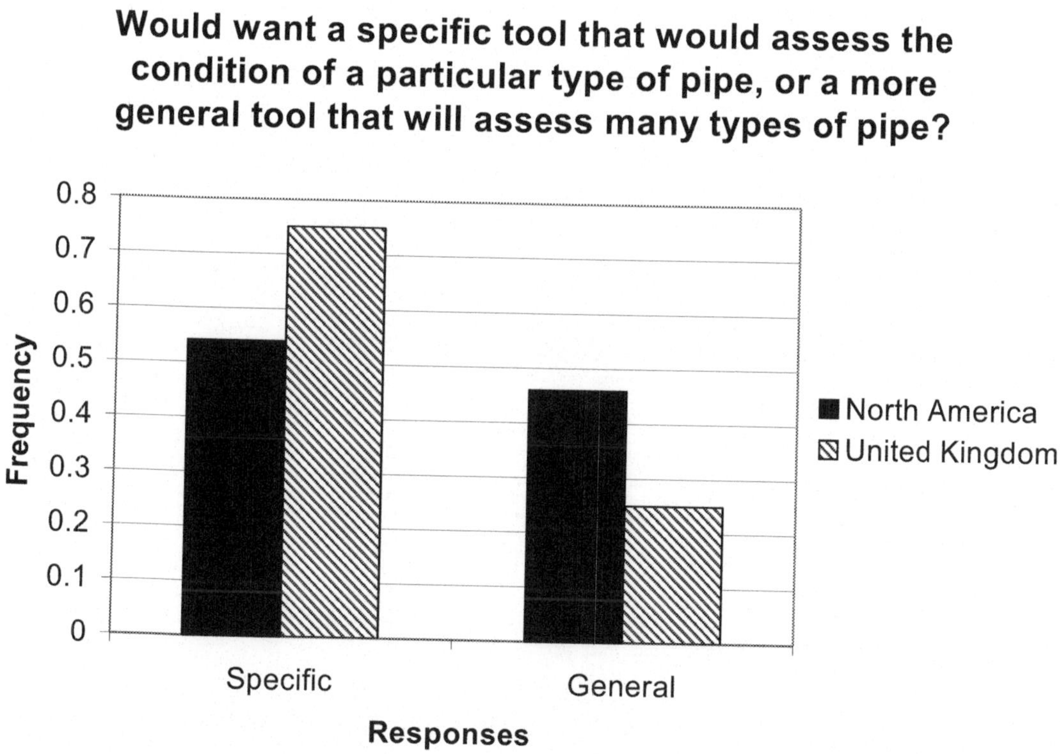

Figure 2.16 Type of tool required

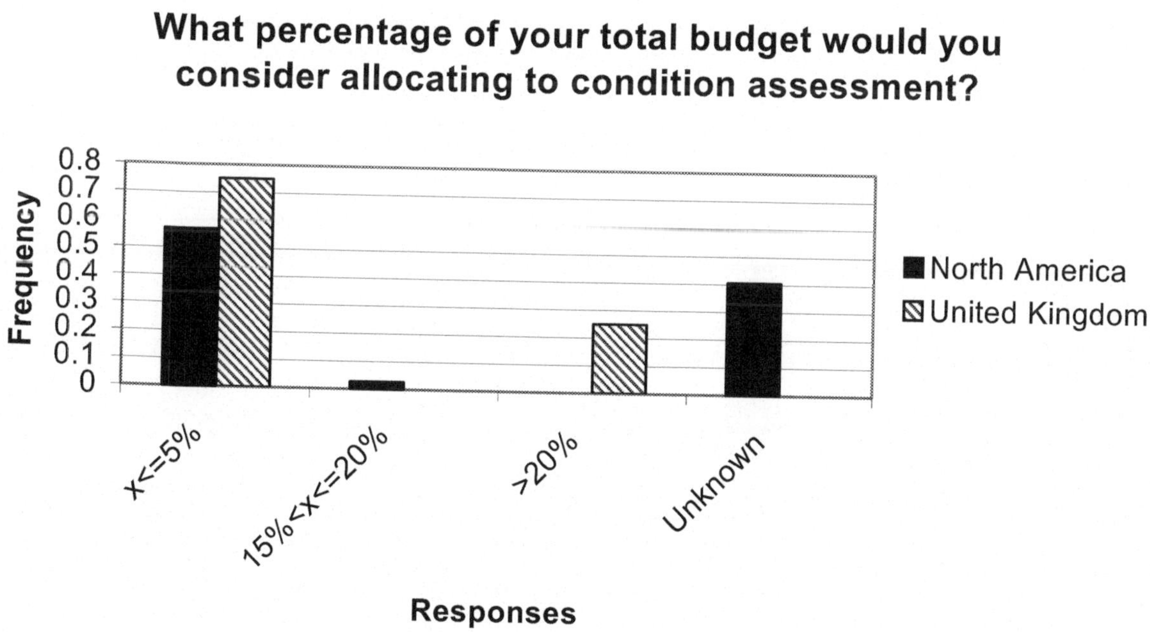

Figure 2.17 Expenditure on condition assessment

CHAPTER 3
TECHNOLOGICAL REVIEW

INTRODUCTION

The status of each technology identified by this review has been assessed from personal knowledge and also by contacting manufacturers, suppliers, and users. The potential for future developments in this area is explored in a detailed literature review of research papers.

This chapter also identifies existing techniques for non-interruptive condition assessment inspection. In addition, those technologies that have the potential for further development also are identified. When available, identified technologies are supported by case studies, which have been compiled in Appendix I. Patent holder information for these technologies is also provided in Appendix I. The technology review has been split into the following four sections:

- Previous AwwaRF projects
- Non-destructive testing techniques
- Identifying and selecting technologies
- Manufacturers and suppliers

PREVIOUS AWWARF PROJECTS

There are a number of AwwaRF projects, complete or ongoing, in the area of distribution system maintenance and renewal related to this project, those with direct relevance are listed below.

- Nondestructive Testing of Water Mains for Physical Integrity [Project #507]. Published in 1992.
- Electromagnetic Inspection of Pre-stressed Concrete Pressure Pipe [Project #2564]. Published in 2001.
- Nondestructive, Noninvasive Assessment of Underground Pipelines [Project #355]. Published in 2002.
- Techniques for Monitoring Structural Behavior of Piping Systems [Project #2612] (To be published in 2003).
- Testing and Condition Assessment of Joints in Water Distribution Pipelines [Project #2689] (To be completed in 2004).

Nondestructive Testing of Water Mains for Physical Integrity [Project #507]

The study reviewed:

- Ultrasonic (US)
- Remote-field eddy current (RFEC)
- Magnetic flux leakage (MFL)
- Acoustic emission (AE)

Steel pipes were the most applicable to these techniques but testing became progressively more difficult for ductile iron, gray cast iron and concrete pressure pipes. However, the presence of a thin cement mortar lining provided a smooth, uniform surface enhancing the ability of the in-pipe inspection tool.

Remote field eddy current technology was found to be the most successful method evaluated, being relatively compact and portable. When data processing technologies are applied to the generated data the results are simplified and easy to understand.

Some of the recommendations based on the study were

- The remote field eddy current technique should be considered for continuous in-line condition assessment of water mains constructed of steel, ductile iron and cast iron pipes. The equipment should be developed further for application to larger pipes and longer inspection assignments.
- It will be necessary to package ultrasonic and magnetic flux leakage inline inspection tools in the required shapes to accommodate water mains.
- Acoustic emissions technology should be further investigated with a specific study of a pipeline constructed from concrete pressure pipes. A considerable effort to calibrate the measurement to pipe conditions will probably be necessary.
- Video camera inspection of pipes may be useful in the future to identify obstructions within the pipe that may damage probes and helps explain the data generated.

Electromagnetic Inspection of Pre-stressed Concrete Pressure Pipe [Project #2564]

- The RFEC/Transformer Coupling (TC) tool is able to detect broken wires anywhere along the length of the pipe and anywhere around the circumference of the pipe.
- The RFEC/TC tool is able to detect and resolve multiple regions of broken wires at different axial locations along the pipe.
- The size of RFEC/TC signal shows a strong and direct relationship with the number of broken wires, which means that distressed pipes can be ranked in order of severity by reliable semi quantitative analysis of the RFEC/TC data.
- The RFEC/TC technique is able to predict the number of broken wires by applying calibration information.
- It is possible to extrapolate tool setup calibration data appropriate to new pipelines.

Although capable of giving a good estimate of where the break occurs along the length of the pipe, the technique can give no information at this time as to the circumferential position of the broken wires.

Break detection sensitivity also changes depending on the region of the pipe in which the break occurs. End breaks are harder to detect, which is further complicated by the fact that end effects due to construction variations can mimic a break signal.

Overlapping signals from different regions may interfere with each other and may also be harder to quantify, although in these regions it is clear from the signal that a large region or multiple breaks must be present.

The project extended to only a few different samples of PCCP, however it has prompted more in depth, ongoing studies.

Nondestructive, Noninvasive Assessment of Underground Pipelines [Project #355]

Pigging tests on metal pipes were found to readily find general wall losses and other volumetric flaws such as severe pitting but no discrete flaws or cracks. This project only included tests on ductile iron, but RFEC methods are successfully used on steel pipes in other industries such as the petrochemical industry.

RFEC pigging in pipes requires removing the pipe from service and flushing when pigging is complete.

Field-testing on concrete/asbestos-cement pipes proved that all the methods used for concrete pipe inspection could identify anomalies in the pipe barrel. However, because no destructive testing could be done, it was not possible to definitively say that the methods are detecting anomalies that related to reduced pipe integrity.

Sonic and impact echo (IE) methods require dewatering but AE requires that the pipe be in service.

Most pig systems have been adapted from the petroleum industry. Unlike water pipelines petroleum lines do not contain sharp bends and are not therefore articulated. Pig manufacturers need to carefully create fully articulated systems that can get around any corner.

In order for a pig to pass into a pipe, the valve must be fully opened. The impact of switching to fully opening valves should be studied to determine if it would outweigh the benefits of condition assessment via pigs.

Techniques for Monitoring Structural Behavior of Piping Systems [Project #2612]

Guided Wave Inspection

This technology has been successfully demonstrated in laboratory and field environment on both cast iron and steel pipes. Features and areas of corrosion have been correctly identified. The equipment is portable and can be used on lined and unlined pipes. It provides first level screening for general pipe condition assessment. Up to 200 ft (60 m) of continuous butt welded steel pipe can be inspected from one location. On other pipe configurations, although only a single pipe spool can be inspected, it is ideal at any hidden/buried interface.

Ultra-Sonic Corrosion Monitoring

These tools are commonly used for spot wall thickness measurements, and are more suitable for internally aggressive environments. Could be used in areas of high potential tuberculation or local erosion (sewage/storm water) but the current system is limited to steel fittings.

SAHARA

This was found to be a technique that can be used to periodically monitor a pipeline for signs of leakage activity. The system has been utilized to survey many pipelines and leaks as low as 1-2 liters/hr have been detected.

Pipe Geometry Pig

Caliper pigs are capable of providing a profile of features on the pipe, detecting some types of joint, measuring internal bore and identifying changes. The prototype tool tested showed there had been little internal corrosion due to tuberculation or cleaning. Different pipe classes had been identified from step changes in bore. Further development work to enhance data interpretation is required.

Metal Loss Pig (RFEC)

Laboratory testing of the prototype RFEC system showed it to be capable of measuring pipe wall thickness and identifying areas of local metal loss. Under operation condition the performance was not satisfactory and further development is required to overcome the problems identified.

Pipeline Route Mapping Pig

Inertial Navigation Systems (INS) are able to map pipeline routes, detect bends and assist with the identification of the directional magnitude of bends. On the prototype tool tested, further characterization of bends is required. Changes in direction across some joints were detected but it was not possible to separate the dynamic effect of the pig. More work is required to determine interaction between pipe position and direction and the pig movement within the pipe

Testing and Condition Assessment of Joints in Water Distribution Pipelines [Project #2689]

The objective of the project is to investigate techniques for testing and condition assessment of joints in water distribution pipelines. The proposed approach is to identify and document key problems associated with failure of joints in pipelines. Additionally, the study will evaluate and report on the performance of existing and emerging techniques for inspection and condition assessment of a wide variety of joints used in water distribution systems. Methods to rehabilitate joints without the expense and disruption of an open excavation will also be investigated.

A key target of the work will be to reduce the costs incurred by water utilities due to leaks or pipe breaks at joints, and to reduce water quality deterioration.

NON-DESTRUCTIVE TESTING TECHNIQUES

Traditional Techniques

There are number of alternatives to physical assessment of the transmission main itself. Alternative techniques such as soil resistivity, soil potential surveys, and hydraulic analysis can give additional information useful to the assessment of pipe condition. For example, a relationship can often be seen between pipeline corrosion and soil resistivity (Randall-Smith, Russell, and Oliphant 1992). A soil resistivity survey can provide additional information to a condition assessment study. Alternative techniques and technologies include the following:

- pipeline history
- location of mains
- pressure testing
- flow tests and Hazen Williams C tests
- soil resistivity
- coupons and samples
- visual inspection

Pipeline History

Throughout the operational life of a transmission pipeline water companies collect large amounts of information relating to pipeline material, date of construction, burial condition, failure history, etc. This historical information can be a valuable source of base data for the engineer undertaking a condition assessment study. It is important that water companies are active to maintain records and present them in a useable form.

Location of Mains

The relationship of a pipeline to the surrounding environment can also provide useful information. Factors such as burial depth, depth of water table, protection by an easement and proximity to heavily traveled roadway present the engineer with data that can be used in the condition assessment and decision making process.

Pressure Testing

During the commissioning phase of constructing a new water transmission pipeline it is usual to undertake a hydrostatic pressure test for identification of leaks. In industries such as oil & gas it is recognized practice to undertake periodic repeat pressure tests in order to meet compliance with regulations. There is always a risk with such procedures that leaks occurring during a test may be in locations where it is difficult to make a repair.

Flow Tests and Hazen-Williams C Test

In water transmission mains tuberculation and debris can lead too much reduced hydraulic performance. There are several formulae including Hazen-Williams that may be applied to the calculation of losses in water pipes. Flow tests together with application of appropriate calculations can be used as a means of quantifying reduction in hydraulic performance of a pipe.

Soil Resistivity

Testing of soil resistivity involves taking two or three measurements from each area under investigation. Four electrodes are inserted in the soil in a straight line directly above the pipe and along its axis. A voltage is impressed between the two outer electrodes and potential drop measured between the two central electrodes. This is used to calculate the soil resistivity (Randall-Smith, Russell, and Oliphant 1992).

Coupons and Samples

The assessment of coupons and pipe samples is common practice by many water companies. This involves the removal of a pipe section, or coupon, for assessment and its substitution by a replacement pipe. The assessment process typically involves measurement of key dimensions, grit blasting to remove corrosion products, measurement of corrosion pit depths and extent of corrosion, and metallographic examination of wedge cut from sample.

Visual Inspection

In large diameter e.g. greater than 54-inch (1.4-m) water transmission mains it may be possible to dewater and gain entry to the inside of the pipe. This procedure can allow for visual inspection of the internal pipe wall to identify coating defects, cracks and water ingress (leaks).

A number of water company delegates attending the workshop provided information relating to their own approach, a number of these are detailed below.

Utility A

Utility A serves approximately 3 million residents through a network totaling some 7,200 miles (11,500 km) of water distribution mains and a further 280 miles (448 km) of water transmission mains. The utility is currently focusing its efforts into determining the condition of transmission mains larger than 24-inch (600-mm) diameter. The methodology for assessing pipeline condition includes pipeline history, location of pipeline, pressure testing, and soil corrosivity reports.

Pipeline history looks at a number of parameters that include age of pipeline, type of material, leak reports, condition of valves, and documentation from visual inspections. Location of pipes again investigates a number of pipeline parameters; these include burial depth, horizontal alignment along roadway, and depth of water table. The second study also asks question such as, is the pipe in an easement and is it located under a heavily traveled roadway?

Pressure testing of transmission pipes comprises of the isolation and pressurization of sections to 150% of the normal working pressure. Pipes are repaired and re-tested to ensure that they meet the requirement. Finally, the utility evaluates this information together with soil corrosivity maps to finalize its assessment.

Utility B

Utility B delivers 130 million gallons of water per day through a 3,000-mile (4,800-km) network of pipes. The Water Works is currently running a program that in 2002 replaced 30 miles (1%) of pipe. Within its decision making process extensive use is made of historical maintenance data, customer complaints information, installation information, CCTV footage of pipe interior, flow tests and Hazen Williams C tests, and removal of coupons for physical assessment.

Results from all of these techniques are used in the decision making process of whether to replace a main, to rehabilitate a main, or simply to do nothing. Utility B has also investigated the use of Thermal Infrared imaging to locate voids and potential leak sites, techniques such as this are seen to be valuable as they require no excavation or depressurization of the transmission main.

Utility C

Utility C has 9,100 ft (2.8 km) of 72-inch (1.83-m) and 78-inch (1.98-m) PCCP delivering water from a major treatment facility. The company has experienced a history of main breaks dating to 1996. It was decided to dewater the main and carry out an internal visual inspection of the entire length. The inspection resulted in identification of three joints requiring further action through excavation and 80 joints requiring rehabilitation by internally grouting.

Conventional Technologies

There are a number of conventional, well-proven non-destructive testing techniques that have been applied to pipeline inspection in the oil, gas, and water industries. The technology review has gathered information on those deemed to be most appropriate for condition assessment inspection of water transmission mains.

Non-destructive testing methods for pipeline condition assessment identified by the review include:

Existing Condition Assessment Inspection Devices:

- Acoustic Emission Monitoring
- Acoustic Leak Detection
- Broadband Electromagnetic
- CCTV
- Long Range Ultrasonic (Guided Wave)
- Magnetic Flux Leakage
- Remote Field Eddy Current
- Remote Field Eddy Current – Transformer Coupling
- Seismic Pulse Echo
- Ultrasonic

Emerging Condition Assessment Inspection Devices:

- Ultrasonic

Applicable Devices from other Industries:

- Inertial Mapping Unit
- Caliper (Geometry)
- Magnetic Flux Leakage
- Ultrasonic

Table 3.1

Acoustic emission monitoring

Assessment technology	Acoustic Emission
Target pipe material(s)	PCCP
Assessment parameter(s)	Wire breaks.
Diameter range (inch)	≥24
Maximum range (feet)	5,280
Inspection rate (ft/min)*	600

*Inspection rate refers to the speed at which data is acquired, it does not include the time to analyze data and report results.

Existing Condition Assessment Inspection Devices

Acoustic Emission Monitoring. Acoustic emission (AE) techniques can be used to monitor pre-stressed concrete cylinder pipes for deterioration. The technique (Table 3.1) employs an array of acoustic sensors that measure the dynamic response caused by the energy released when pre-stressing wires fail.

The technique (Figure 3.1) employs a hydrophone array that is installed into the pipe under pressure. The array remains in place for the duration of the monitoring period; wire failures are captured by the system and reported to the client. Table 3.2 lists the advantages and limitations of this technique.

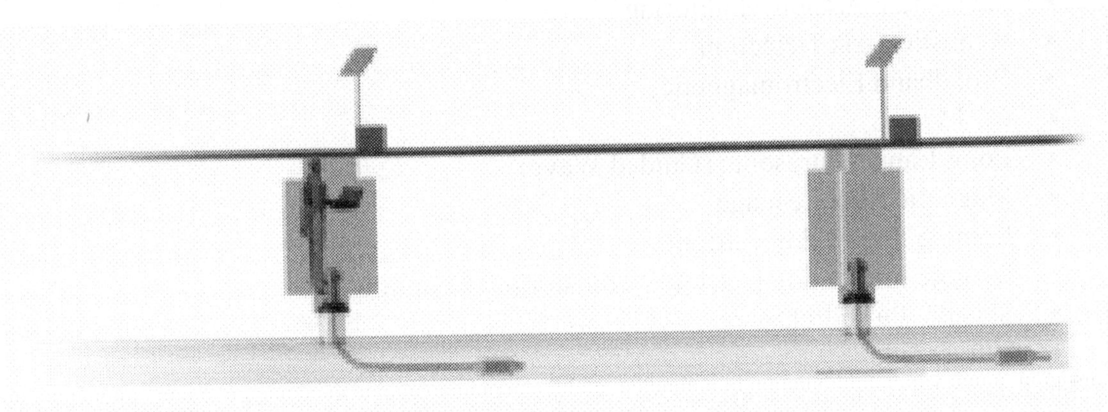

Figure 3.1 Schematic showing insertion of SoundPrint® hydrophone array into water main (*Source*: SoundPrint, SoundPrint® Infrastructure Management Technology, 2002)

Table 3.2

Advantages and limitations of acoustic emission technique

Advantages	Limitations
Inserted into water main under pressure, no interruption to supply	Does not provide a baseline measurement, only monitors
Web enabled reporting system provides event information	

Table 3.3

Acoustic leak detection

Assessment technology	Acoustic leak detection
Target pipe material(s)	Any
Assessment parameter(s)	Leakage
Diameter range (inch)	>12
Maximum range (feet)	6500
Inspection rate (m/s)	0.2

Acoustic Leak Detection. In-pipe, acoustic leak detection systems (Table 3.3) can locate leaks in large diameter (≥12-inch) water mains. These employ a sensor head, which is attached to an umbilical and inserted into the main through a small diameter access fitting. The sensor head is carried by the flow of water and has a range of up to 6,500 ft (2 km). Insertion into the main can be through any on line fitting with a bore ≥2 inches (50 mm). The schematic shown in Figure 3.2 illustrates the configuration of hardware for an acoustic leak detection system. Table 3.4 lists the advantages and limitations of acoustic leak detection.

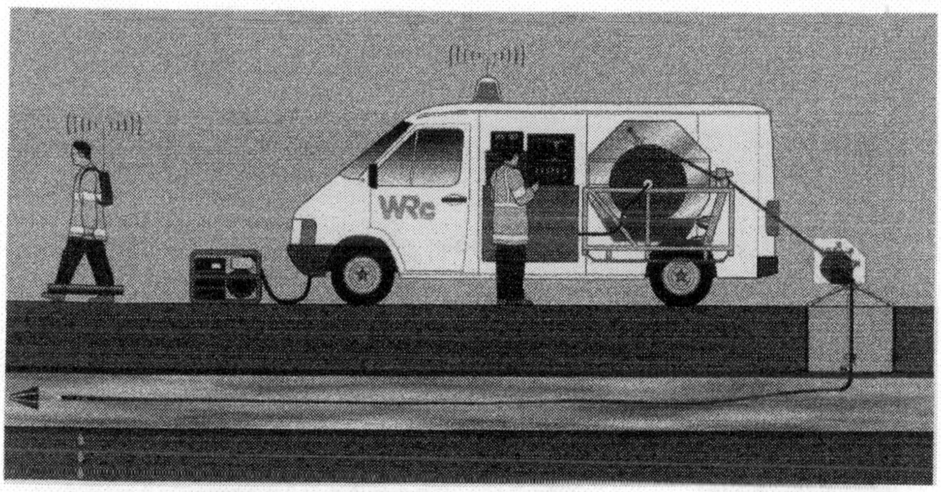

Figure 3.2 Schematic showing configuration of hardware for Acoustic Leak Detection (*Source*: WRc 2003)

Table 3.4

Advantages and limitations of acoustic leak detection

Advantages	Limitations
Avoids disruption of the water supply to the customer	Inspection range can be limited by pipeline geometry (number of bends)
Wide diameter range available	
Requires only a small diameter access point	

Broadband Electromagnetic. Broadband electromagnetic techniques can be used to assess ferrous pipe wall condition and locate illegal tap-ins in water and sewer mains (Table 3.5). Tools are available for external and internal use.

Pipe wall condition assessment is by means of an internal condition assessment probe; this allows continuous data to be recorded along extensive lengths of pipeline (Figure 3.3). The condition assessment probe can be winched or rodded through de-pressurized pipes. Results of wall thickness are reported graphically or as color contour plots. Table 3.6 lists advantages and limitations of this technique.

Table 3.5

Broadband electromagnetic

Assessment technology	Broadband Electromagnetic
Target pipe material(s)	Steel, cast iron, ductile iron
Assessment parameter(s)	Remaining wall thickness
Diameter range (inch)	≥3
Maximum range (feet)	2,000
Inspection rate (ft/min)	9 to 11

Figure 3.3 Picture of Broadband Electromagnetic internal condition assessment tool (*Source*: Rock Solid Group Pty. Ltd., *Pipeline Condition Assessment Techniques for Ferrous Pipelines*, Roubal 2003).

Table 3.6

Advantages and limitations of broadband electromagnetic technique

Advantages	Limitations
Able to survey through external coating and internal linings No upper limit on pipe diameter	Requires pipe to be depressurized during the assessment and full bore access Inspection rate is only a few feet per day in large diameters

Optical – Closed Circuit Television (CCTV). Equipment for inspection of pipelines has been commercially available since the early 1970s. In larger pipe diameters (greater than 12 inches), inspection ranges up to 1,300 ft (400 m) are possible. Camera technology (Table 3.7) has become increasingly sophisticated and many systems now offer color, zoom, pan and rotate facilities. A picture of a CCTV internal assessment tool is shown in Figure 3.4. Table 3.8 displays the advantage and limitation of this technique.

Table 3.7

Optical – closed circuit television

Assessment technology	Optical
Target pipe material(s)	All
Assessment parameter(s)	Tuberculation, lining, and blockages
Diameter range (inch)	≥4
Maximum range (feet)	1,300
Inspection rate (ft/min)	Not available

Figure 3.4 Picture of CCTV internal assessment tool (NWTC)

Table 3.8

Advantage and limitation of optical technique

Advantage	Limitation
Can detect visible internal damage, lining deterioration, and partial blockages	Only allows a subjective assessment of condition

Long Range Ultrasonic (Guided Wave). Externally operated low frequency long-range ultrasonic testing (LRUT) techniques can be used for the detection of both internal and external corrosion (Table 3.9). The principal advantage of this technique is that it can assess up to 100 ft (30.5 m) of pipe in either direction from a single test point.

The technique does not provide a direct measurement of pipe wall thickness; but is sensitive to a combination of depth and circumferential extent of any metal loss. Detectable flaw area is presented as a percentage of pipe wall cross section. This is an average volume of metal loss technique, so you could still possibly get a failure at a weak point. One tool used in this technique is shown in Figure 3.5. Advantages and limitations of this technique are listed in Table 3.10.

Table 3.9

Long range ultrasonic

Assessment technology	LRUT technique
Target pipe material(s)	Steel
Assessment parameter(s)	Pipe wall cross section
Diameter range (inch)	2 to 48
Maximum range (feet)	100
Inspection rate (ft/min)	Not available

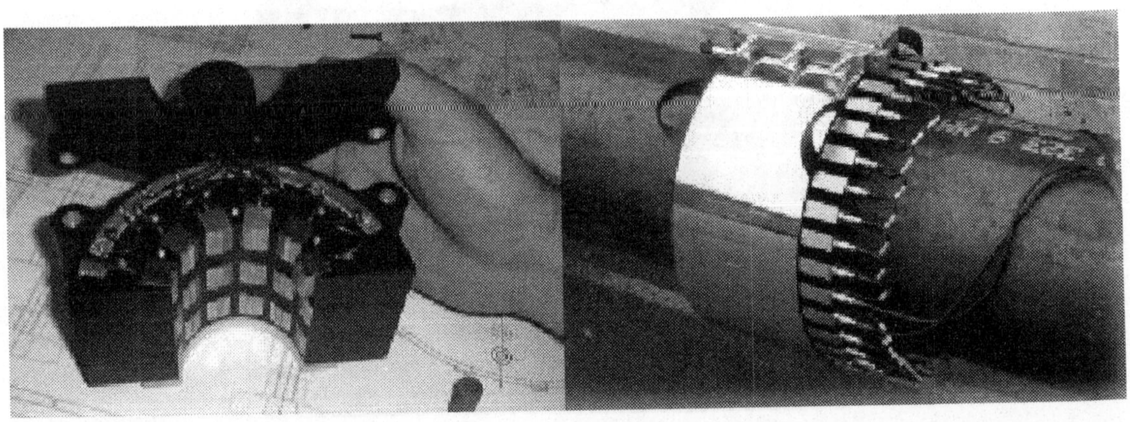

Figure 3.5 Picture of Teletest tool (*Source*: Lebsack, S., Non-Invasive Inspection Method for Unpiggable Pipeline Sections, 2002)

Table 3.10

Advantages and limitations of long range ultrasonic technique

Advantage	Limitations
No interruption to supply, non-intrusive technique	Can only be used where pipe is exposed or by digging access pits
	Only reports reduction in pipe wall cross section, not deepest pit
	Could miss a critical defect
	Only proven on welded steel pipelines

Magnetic Flux Leakage. Magnetic Flux Leakage (MFL) is commonly used as an in pipe inspection technique in the oil and gas industry, but is also used as an external inspection technique for pipes, including water pipes (Table 3.11).

MFL detects defects in ferrous materials by applying a saturating magnetic field to the material using large magnets. Figure 3.6 shows a schematic of the technique applied inside a pipe.

Table 3.11

Magnetic flux leakage

Assessment technology	Magnetic Flux Leakage
Target pipe material(s)	Cast Iron, Ductile Iron, Steel
Assessment parameter(s)	Wall thickness
Diameter range (inch)	>3
Maximum range (feet)	Not available
Inspection rate (ft/min)	Not available

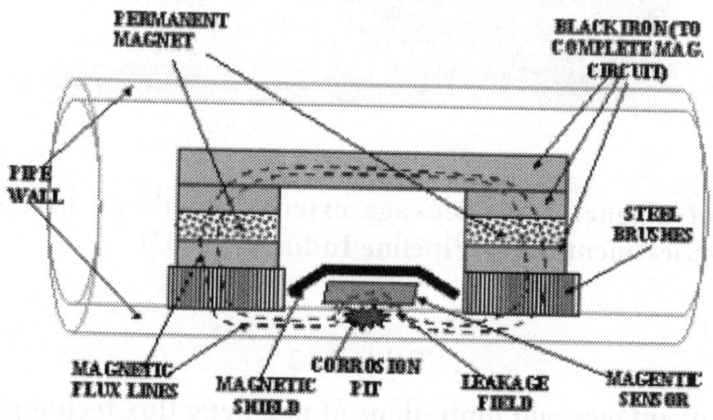

Figure 3.6 Schematic showing magnetic flux leakage technique (Source: CSIRO Manufacturing and Infrastructure Technology, 2004)

If defects are present in the pipe, the magnetic flux 'leaks' and the change is noted by detector coils on the inspection tool. MFL response to pipeline anomalies depends primarily upon the magnetic properties of the pipeline material and the geometry of any defects. Any material that has been removed can be detected as it reduces the ability of the pipe to carry magnetic flux. In steel pipes, mechanical damage can also be detected as the magnetic properties of steel are changed locally through plastic deformation (Bickerstaff, Vaughn, Stoker, Hassard, and Garrett 2002). Traditional MFL is normally limited to a wall thickness of around ½ inch (10-12 mm). MFL is an indirect method of inspection that requires calibration of its signals to provide useful data. MFL inspection is also dependent upon magnet strength, inspection speed and the stress factor on the pipe wall (Plotnikov and Clapham 2002).

Non-intrusive, externally operated magnetic flux leakage tools (,) are currently available for water pipeline condition assessment. On small diameter pipe up to 6 inches (150 mm), tools are available that fully encircle the pipe. On larger diameter pipes, tools are used to inspect segments of the pipe. The tools are traversed along the pipe and inspect segments of the pipe as it travels along the track. The tool is capable of being used on steel, ductile iron and cast iron pipes. These tools require clear access to the outside of the pipe, which may involve excavation. Table 3.12 shows the advantage and limitation of this technique.

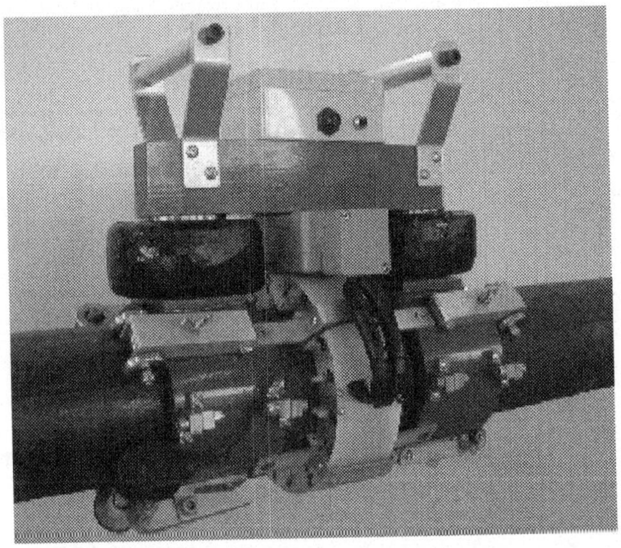

Figure 3.7 Picture of magnetic flux leakage external condition assessment tool (*Source:* Wayman, M., Risk Assessment in the Pipeline Industry, 2003)

Table 3.12

Advantages and limitations of magnetic flux technique

Advantage	Limitation
No interruption to supply, non-intrusive technique	Can only be used where pipe is exposed or by digging access pits

Remote Field Eddy Current (RFEC). Cast and ductile iron pipes account for approximately 70% of the North American water distribution network (Austin 1997). Historically, inspection methods for these have lagged behind those for other pipe materials primarily due to the nature of most water distribution systems (short pipe sections, 90 degree bends and valves that do not open fully). Remote Field Eddy Current (RFEC) techniques have come to the fore recently however as a method of internal inspection that can detect internal corrosion and overall wall loss (Table 3.13).

The technique involves the creation of an electromagnetic field, using low frequency Alternating Current (typically 20-2000 Hertz), inside a pipe using a solenoidal exciter coil (Atherton 2000). Two distinct paths exist for the field between the exciter and the detector array. One part of the field travels along the inside surface of the pipe. This is attenuated rapidly by circumferential eddy currents induced in the pipe's wall. This is the Direct Field. The remainder of the field propagates radially through the pipe wall. When it reaches the outer wall, the field travels along it with little further attenuation. After a short distance (typically 2-3 times the pipe diameter) this field will re-diffuse into the pipe where it is now stronger than the direct field and will be detected by the detector coil (AMG 2002). This portion is referred to as the Remote Field. The total magnetic field and eddy current flow at any given point is the combination of the remote and directly coupled fields.

To detect the field, one or more detector coils are placed at least 2.5 pipe diameters behind the exciter coil close to the inside wall of the pipe. At this distance, the remote field is stronger than the directly coupled field due to the different rates at which the two fields decay. The behavior of the electromagnetic field is dependant on the thickness of the pipe wall and therefore can be used to detect metal loss, corrosion pitting and axial cracking (Battelle 2002). Anomalies anywhere within the remote path cause changes to the magnitude and phase of the received signal and can therefore be used for detection of defects (AMG 2002). See Figure 3.8.

RFEC is currently the only method of continuous inspection available, in the water industry, for determining the condition of cast and ductile iron pipes; it may also be used to inspect steel pipes. The current primary use for RFEC is in the inspection of small diameter pipelines (Makar and Kleiner 2000) such as heat exchanger and pressure tubes.

Unlike conventional Eddy Current techniques, RFEC is capable of inspecting the entire wall thickness of a pipeline without needing to use ultra low frequency. It is hoped therefore that RFEC might provide a solution to the problem of detecting Stress Corrosion Cracking (SCC) (AMG 2002). Traditionally however, RFEC has used low-frequency exciter coils. This typically limits the maximum speed at which an inspection vehicle can travel to around one mile per hour, thereby limiting the potential uses of RFEC as an in-line inspection (ILI) technology.

Table 3.13

Remote field eddy current

Assessment technology	RFEC
Target pipe material(s)	Steel, cast iron, ductile iron
Diameter range (inch)	6, 8, 10, 12, 15, 16 and 18
Maximum range (feet)	3,000
Inspection rate (ft/min)	20

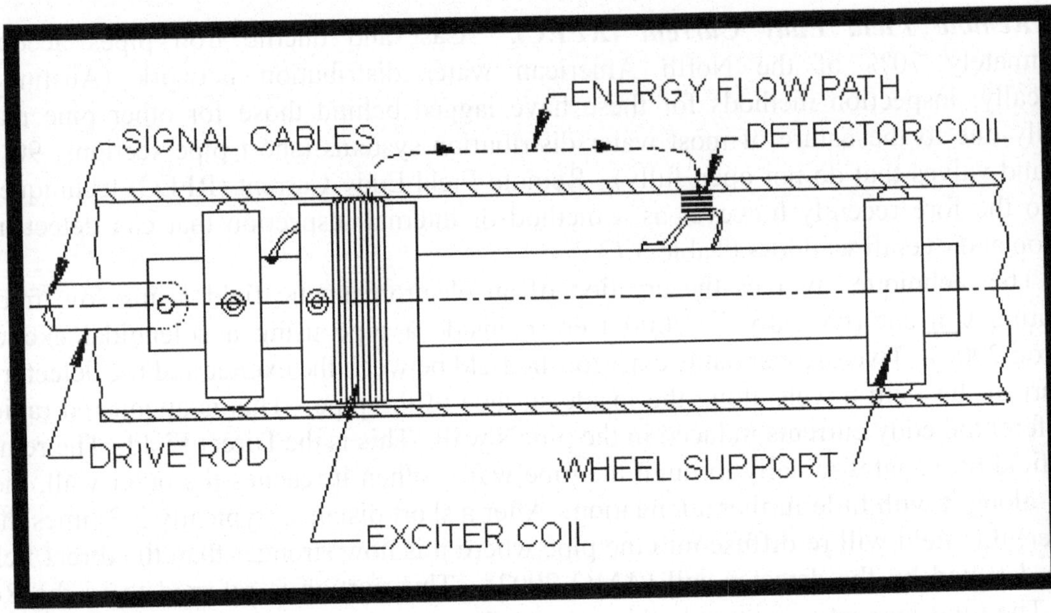

Figure 3.8 Cross section of RFEC ILI tool (*Source*: AMG, Remote Field Through Wall Inspection Technique, 2002)

Existing RFEC systems are deployed in-pipe and propelled through the pipe by water pressure or by winching the tool. A trailing wire line cable connects the tool to a computer above ground. Access to the water main can be achieved through modified fire hydrants or through a tee inserted into the main. Systems are capable of negotiating 90° elbows and passing tees. Advantages and limitations of RFEC technique are shown in Table 3.14.

Remote Field Eddy Current – Transformer Coupling. This technique is used to inspect Pre-stressed Concrete Cylinder Pipe (PCCP). PCCP is a composite structure of high strength steel windings embedded in concrete, which has been used extensively for the high-pressure supply of water to large cities (Table 3.15). There are two types of PCCP. Lined Cylinder Pipe (LCP) has its steel windings wound tightly onto the outer surface of the cylinder. Embedded Cylinder Pipe (ECP) has a layer of concrete between the cylinder and the pre-stressing windings. Both types of PCCP then have an outer coating of mortar applied over the windings to help

Table 3.14

Advantages and limitations of RFEC technique

Advantages	Limitations
Established technology dates back to early 1990's	Pipe requires cleaning prior to inspection
Available to suit a range of pipe diameters	Cost of removing material dislodged by cleaning is high
100% examination of pipe wall	Necessary to discharge water to the environment during survey
Assessment possible in wet or dry pipe	

Table 3.15

Remote field eddy current – transformer coupling

Assessment technology	RFEC/TC
Target pipe material(s)	PCCP
Assessment parameter(s)	Wire Breaks
Diameter range (inch)	16 to 256
Maximum range (feet)	Not available
Inspection rate (ft/min)	Not available

protect them from corrosion. PCCP's may range in size from approximately 16 inches (400 mm) to more than 16 ft (5 metres). Pipes of more than 47 in. (1.2 metres) are generally constructed as ECP's (Mergelas and Atherton 1999).

RFEC/TC is used to detect broken pre-stressing wires in embedded cylinder pipes. These wires are used to maintain compression in the concrete under maximum operating conditions. Corrosion of the steel windings can lead to wire breaks and if the number of broken wires passes a critical point, the concrete will no longer be under compression and a catastrophic failure may occur (Mergelas and Atherton 1999).

RFEC/TC is a new technique that has been developed by the Pressure Pipe Inspection Company (Ontario, Canada) in collaboration with the Applied Magnetics Group (AMG) at Queen's University (Ontario, Canada). See Figure 3.9.

The main difference between this technique and standard RFEC is that RFEC/TC has been optimized for inspection of the solenoidal pre-stressing winding. This behaves similarly to a shorted transformer coil, enhancing the external field coupling when compared to that of a uniform metallic cylinder (May and Atherton 2002). The electromagnetic field produced in this technique interacts with broken pre-stressing wires. Wire breaks interrupt the flow of energy

Figure 3.9 Picture of Pressure Pipe Inspection Company's RFEC/TC tool (*Source*: The Pressure Pipe Inspection Company, Remote Field Study Current/Transformer Coupling, 2002).

Table 3.16

Advantages and limitations of RFEC/TC technique

Advantages	Limitations
Detects single or multiple pre-stressing wires	Requires dewatering of pipe
	Requires human entry into pipes

through the pipe wall, changing the measured field, allowing for detection of broken wires (Makar and Kleiner 2000). The technique has the ability to detect broken windings anywhere along the length or around the circumference of a pipe. It also has the ability to predict the number of broken wires through the application of calibration data (Mergelas and Kong 2001). Even without calibration however, it would be possible to analyse the data semi-quantitatively, identifying those pipelines with most damage (Mergelas and Atherton 1999).

May and Atherton (2002) have carried out experiments on the use of multi-frequency RFEC/TC to identify PCCP wire breaks in the presence of large background field variations as caused by pipe joints. In theory, multi-frequency data contains more information than single-frequency data, and this can be used to improve the accuracy of characterization. Whilst this testing was found to facilitate classification of different types of anomalies, complete separation of wire break responses from joint region responses was not possible. For most practical applications, accurate reference samples would be required to calibrate the instrumentation prior to inspection. The suggestion is therefore that such a technique would be of limited use for the detection and quantification of anomalous wire break regions in PCCP.

RFEC/TC technology can be used to identify single and multiple broken wires providing an assessment of deterioration of concrete pressure pipes.

Table 3.16 identifies the advantages and limitations of the RFEC/TC technique.

Seismic Pulse Echo. Seismic Pulse Echo is the name given to a sonic/ultrasonic technique for evaluating the condition of prestressed concrete cylinder pipe (Table 3.17). The technique can be used to determine pipe flaws such as wire failures, delamination and cracks. The technique was previously identified by AwwaRF project #355.

Table 3.17

Seismic pulse echo

Assessment technology	Seismic pulse echo
Target pipe material(s)	PCCP
Assessment parameter(s)	Wire breaks, delamination, and cracks
Diameter range (inch)	≥54
Maximum range (feet)	Not available
Inspection rate (ft/min)	4

Table 3.18

Advantages and limitations of seismic pulse echo technique

Advantages	Limitations
Successfully identifies defects in PCCP	Requires dewatering and man entry into the pipe
Can be used to determine baseline prior to use of online monitoring techniques such as Acoustic Emission monitoring	Is not non-interruptive
	Not currently available in diameters below 54 inch (1,400 mm)

The Seismic Pulse Echo technique uses impact from a metal sphere to generate ultrasonic shear and compression waves in the pipe wall; a sensor array is then used to pick up the resulting signals. Sound waves in pipe in poor condition travel slower with shear waves, having less amplitude and reflections occurring at later times. Frequency domain characteristics of pipe in poor condition differ from the frequency domain characteristics of pipe in good condition.

The testing procedure can inspect about 1,500 feet (460 m) per day in a large clean pipe. The technique requires dewatering and man entry into the pipe and is therefore limited to diameters above 54 inch (1,400 mm). The advantages and limitations of the seismic pulse echo technique are identified in Table 3.18.

Ultrasonic. Using a probe the ultrasonic inspection technique transmits mechanical waves of short wavelength and high frequency into the pipe wall. A receiver monitors the transmitted wave and any attenuation caused by cracks, cavities, or other discontinuities. See Table 3.19.

Commercial ultrasonic inspection technology is available and is commonly used for external inspection of steel oil and gas pipelines. Ultrasonic probes are also used for wall thickness measurement as a means of assessing the condition of water trunk mains. A methodology for applying this technique is described in the WRc Guidance Manual for the Structural Condition Assessment of Trunk Mains.

Table 3.20 describes the advantage and disadvantage to the ultrasonic technique.

Table 3.19

Ultrasonic

Assessment technology	Ultrasonic
Target pipe material(s)	Steel, cast iron, and ductile iron
Assessment parameter(s)	Wall thickness
Diameter range (inch)	All
Maximum range (feet)	Not available
Inspection rate (ft/min)	Not available

Table 3.20

Advantage and limitation of ultrasonic technique

Advantage	Limitation
No interruption to supply, non-intrusive technique	Can only be used where pipe is exposed or by digging access pits

Emerging Condition Assessment Inspection Devices

These devices are currently under development for water industry application.

Ultrasonic. A group of U.K. companies is currently undertaking a partnership programme of studies evaluating ultrasonic transducers for water pipeline condition assessment. Completed activities to date include:

- Design and manufacture of a range of ultrasonic transducers and associated electronic drive and data recording systems, optimized for the internal examination of water filled cast iron pipes.
- Controlled laboratory based performance tests of individual transducers to determine characteristic responses to water pipeline features and materials.
- Grouping of sensors in various mounting configurations and gathering of data by traversing these through water filled sections of pipe.
- Gathering data from a variety of pipe materials including cast iron, ductile iron, steel, polyethylene, PVC and asbestos cement, and from machined defect spools.
- Development of software algorithms for processing of raw data from the ultrasonic transducers and subsequent extraction of pipe condition related information.

Results to date have demonstrated that ultrasonic sensing technology can be successfully used to detect and measure water pipeline features, as detailed in Table 3.21.

Table 3.21

Water pipeline feature detection and measurement capabilities

Feature	Ultrasonic Transducer Performance
Pipe geometry	Sensors can be used to measure pipe parameters including, inside and outside diameters, wall thickness, eccentricity and ovality
Cracks	Detect both circumferential and longitudinal cracks
Corrosion	Detect corrosion interfaces in cast iron pipes and loss of section in all materials
Tuberculation	Ultrasonic sensors can provide a measurement of the severity and distribution of tuberculation in the pipe
Lining	Detect whether or not a lining is present
Joints and fittings	Processed data can be used to detect the signatures of joints and fittings
Pipe material Identification	Processed data can be used to identify pipe material

It is expected that further development of data processing algorithms will allow more accurate sizing of cracks and quantification of corrosion extent.

Applicable Devices from other Industries

The following in-pipe systems are commonly used in the oil, gas, and petrochemical industries but with potential to be re-engineered for use in water mains.

Inertial Mapping Unit. Inertial Navigation Systems use highly accurate angular rate sensors, gyroscopes, and accelerometers. Inertial Navigation Systems are used to determine the dynamic properties and trajectory of a moving object. They are also used for navigation, guidance, and control or stabilization of objects. The translational and rotational movements and velocities are detected by using accelerometers and gyroscopes (Figure 3.10).

By using signal processing algorithms and compensating for the Earth's rotational vector and gravitational acceleration, kinematic quantities such as position, orientation, velocity, acceleration, etc. can be determined in three dimensions and in real time.

There are numerous types of sensor that can be used depending on the end users needs. These sensors include all types of ring laser gyros (RLG), fiber-optic gyros (FOG), piezo-vibrating gyros (PVG), dynamically tuned gyros (DTG), as well as micromachined sensors (MEMS) and inertial-grade servo accelerometers (iMAR GmBH 2003).

Operators of oil and gas pipelines use inertial Navigation Systems to map the location of their assets in three-dimensional space.

A more recent development has been to use an IMU to provide as laid route information for cable ducts (DuctRunner, http://www.Reduct.net).

Caliper (Geometry). In the oil and gas industry, in pipe geometry or mechanical, caliper tools use instrumented fingers or mechanical caliper arms to detect and measure mechanical defects. It is possible to measure internal diameter changes and to detect dents, buckles, welds, joints, offtakes and other fittings. A caliper pig is shown in Figure 3.11.

Magnetic Flux Leakage. The principles of MFL inspection have been described previously. A number of companies offer in-pipe inspection services to the oil and gas industries. These services have been used in these industries for approximately 25 years. Figure 3.12 shows an example of a MFL pig.

Figure 3.10 Example of an inertial mapping pig from the PII Pipeline Solutions range (*Source*: PII Group Limited, 2002)

Figure 3.11 High Resolution 32 Sensor caliper pig (*Source*: Profile International Ltd, 2002)

Figure 3.12 BJ Pipeline Inspection Services Vectra MFL Tool (*Source*: BJ Services Co.)

Companies offering these services have a wealth of experience in the pipeline inspection industry. The technology is well established for inspection of high-pressure steel pipelines. However, there is currently no capability for in-pipe inspection of water mains using the MFL technique.

Ultrasonic. A number of companies offer in-pipe inspection services using ultrasonic technology. In-pipe ultrasonic systems are capable of measuring wall thickness and corrosion depth, and some systems have the ability to detect cracks. The systems are designed for operation in steel pipelines and there is currently no capability for in-pipe inspection of water mains. Figure 3.13 shows two examples of these devices.

Figure 3.13 Pictures of ultrasonic tools from the PII Pipeline Solutions range (*Source*: PII Group Limited, 2002)

IDENTIFYING AND SELECTING TECHNOLOGIES

This section will look at the technologies identified by thIS Technology Review and compare them with business needs. Table 3.22 forms the basis for identifying needs not addressed by equipment and services from manufacturers and suppliers in this chapter.

Table 3.22

Summary of technologies versus assessment parameters

Assessment Technology	Target pipe material(s)	Diameter range (inch)	Maximum range (feet)	Inspection rate (ft/min)	Interruption to supply	Pre-inspection cleaning	No special access	Survey through coatings	External/Internal assessment
Acoustic Emission	PCCP	≥24	5,280	–	N	N	Y	–	I/E
Acoustic Leak Detection	All	>12	6,500	40	N	N	Y	–	I
Broadband Electromagnetic	CI, S, DI	≥3	2,000	9-11	Y*	N	Y*	Y	I/E
CCTV	All	≥4							
Long Range Ultrasonic	S	2-48	100	–	N	N	N	Y	E
Magnetic Flux Leakage	CI, S, DI	>3	–	–	N	N†	N	N	E
Remote Field Eddy Current	CI, S, DI	6-18	3,000	20	Y	Y‡	N	Y	I
Remote Field Eddy Current - TC	PCCP								
Seismic Pulse Echo	PCCP	>54	–	4	Y	N	Y	Y	I
Ultrasonic	CI, S, DI	All	–	–	N	Y†	N	Y	E

* Internal assessment requires full bore access and interruption to supply

† Limited cleaning of external pipe face to remove dirt and loose corrosion product

‡ Requires light cleaning with a foam swab

The previous sections brought together requirements for a non-interruptive condition assessment inspection device developed through the literature survey and project workshop. In this section requirements are analyzed against capabilities of available technologies.

The technique types considered within this analysis are as follows:

where AE = Acoustic Emission
AL = Acoustic Leak Detection
BE = Broadband Electromagnetic
LU = Long Range Ultrasonic
MF = Magnetic Flux Leakage
RE = Remote Field Eddy Current
SP = Seismic Pulse Echo

TR = Remote Field Eddy Current Transformer Coupling
TV = CCTV
UT = Ultrasonic

Issues to be considered in the analysis of existing condition assessment inspection devices for water transmission mains include the operational capability of each tool in the following areas (see Figure 3.14). The figure has been compiled along the lines of a selection procedure or 'road map'; in the future it could form the basis of a decision support process for water pipeline engineers.

- Pipe material
- Pipe diameter
- Interruption to supply
- Partial/Full assessment
- Pre-inspection cleaning
- No special access
- Survey through coatings
- Maximum range
- Inspection rate

MANUFACTURERS AND SUPPLIERS

A list of manufacturers and suppliers can be found in Appendix J.

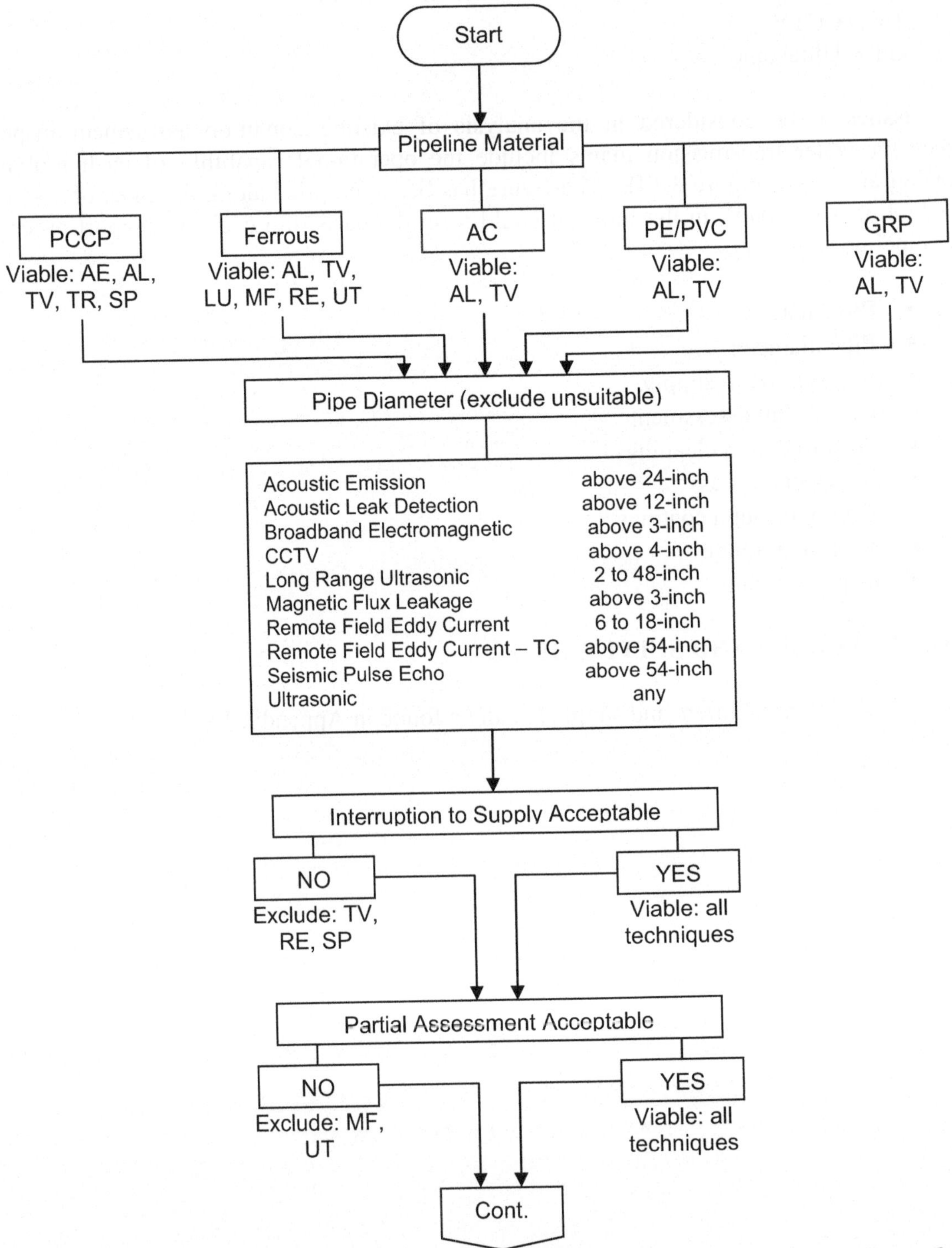

Figure 3.14 Draft selection procedures for condition assessment inspection of water transmission mains (*Source:* PII Group Limited, 2002.)

(continued)

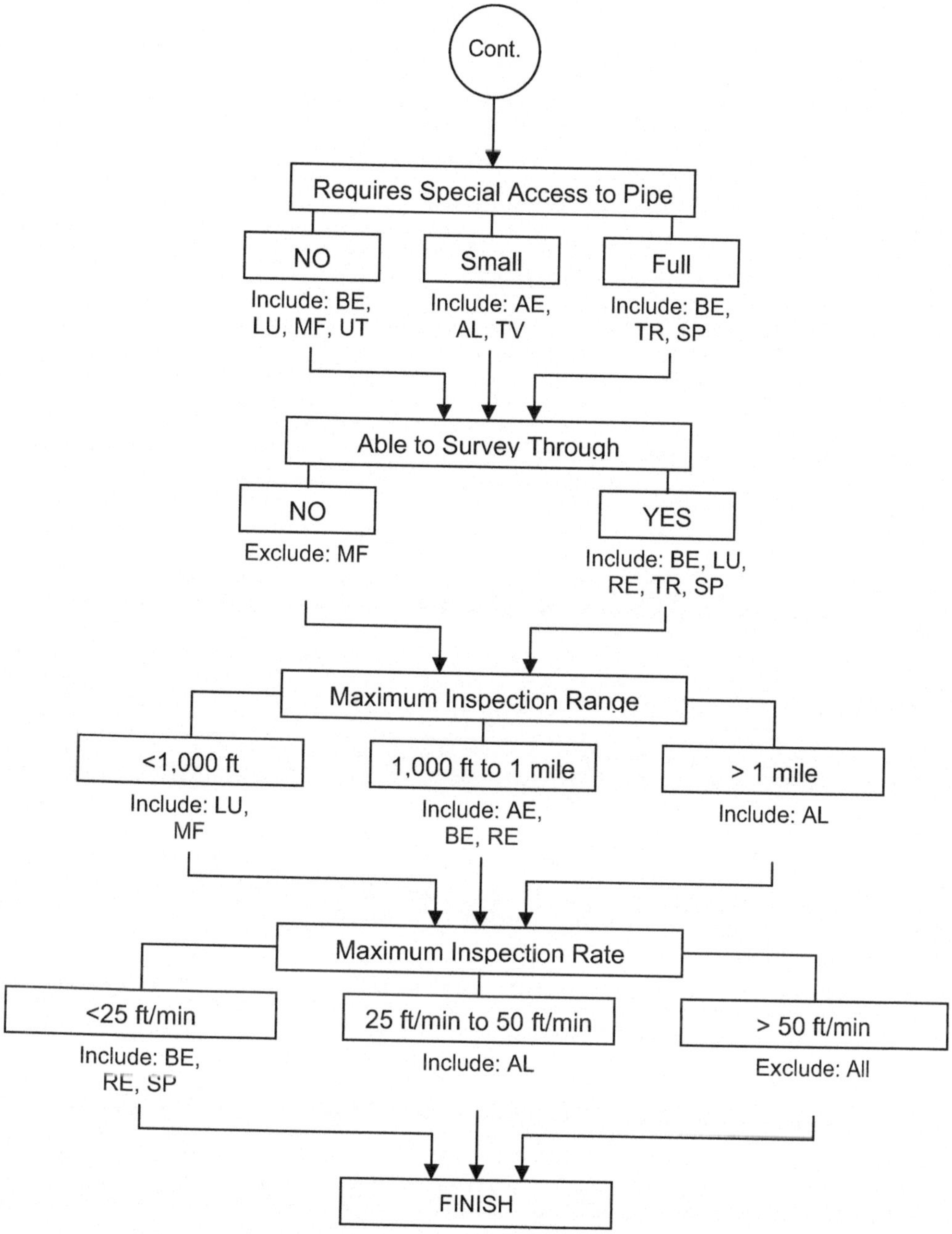

Figure 3.14 (Continued) Draft selection procedures for condition assessment inspection of water transmission mains (*Source:* PII Group Limited, 2002)

CHAPTER 4
NON-INTERRUPTIVE CONDITION ASSESSMENT REQUIREMENTS

INTRODUCTION

After reviewing the business needs and the information gained through the technology survey and the workshop, this section considers the requirements for non-interruptive condition assessment.

TARGET APPLICATION

The case for condition assessment of large-diameter transmission mains is a strong one. Transmission pipelines are considerably more expensive to purchase and install than distribution mains. Should a transmission main fail the results can be catastrophic and very expensive to remedy due to the high volume and high pressure of the water contained within the pipe. Transmission mains tend to be very critical as these supply clean water for a large number of homes, businesses, hospitals and industrial processes. So it is vital that the levels of service be maintained, as the consequences could be high.

How a water company utilizes condition assessment will vary from company to company depending upon their own needs and individual strategies however, individual mains will be prioritized by the risk they pose. Factors typically considered in assessing risk may include probability of failure (e.g. pipe condition, pipe deterioration rate, likelihood of excessive loading and failure history) and consequences of failure (e.g. the number, type, degree of effects on water supply, customers, surroundings (e.g. infrastructure, traffic, businesses, environment) and budgets). This risk assessment would allow areas of the strategic mains network to be targeted for further condition assessment.

GENERAL CONDITION ASSESSMENT REQUIREMENTS

What does a water utility owner or operator want to know about its pipe network? This is the question that the non-destructive evaluation industry has to be able to answer if it wants to break into what at the moment is a relatively untapped market. The water industry would like to know and understand what the tools and services associated with condition assessment can actually do for them. The output from the workshop suggested that water utilities would, in general, like to be able to:

- Determine the probability of large diameter pipe failures.
- Determine the location of potential pipe failures.
- Monitor thrust restraint in larger diameter pipes.
- Assess the condition of pipe barrels, joints, loading and the rate of deterioration.
- Identify loading regimes that may lead to subsequent pipe failure.

Conversely, the condition assessment technology providers would like to know how much the utilities are willing to pay per mile per year for inspection, how much they are willing to invest up front and how long utilities are willing to wait to reach the break even point. Other elements such as what is an acceptable cost/benefit ratio for utilities to justify investing in a new inspection device or service, the required time and location accuracy of assessments, remaining service life estimates, also need to be considered.

Outputs Required

The outputs required from any condition assessment process or tool will vary depending upon the components within the particular network, but several general parameters (covering all material types), outlined during the workshop, were as follows:

- Leaks, cracks, loading conditions, joint condition, number of wire breaks and deterioration rate.
- Condition of lining (although it is of a lower priority).
- Detection of illegal off-takes and connections.

Some of these parameters are more applicable to certain pipe material types than others, and a summary of the potential failure mechanisms, which would need to be assessed for each of the various material types, is given in Table 4.1.

Table 4.1

Potential pipe failure mechanisms

Mechanism	Metallic			Mineral based			Plastic		
	Cast iron	DCI	Steel	PSC	PCCP	AC	PVC	PE	GRP
Internal corrosion (lining)	X	X	X						
External corrosion (coating)	X	X	X		X				
Wall thinning	X	X	X						
Through-wall cracks	X	X	X	X	X	X	X	X	
Surface cracks	X	X	X	X	X	X	X	X	X
Wire breaks					X				
Delamination					X				X
Level of pre-stress	X	X	X	X	X	X			
Softening				X	X	X	X	X	
Leaks	X	X	X	X	X	X	X	X	X
Geometry			X				X	X	X
Pipe bedding			X	X	X	X	X	X	X
Joint pull-out	X	X		X		X	X		X
Socket/spigot cracking	X	X		X	X	X	X		X
Joint misalignment	X	X		X		X	X		X

Note: Tuberculation only gives an indication that corrosion may be occurring

Spatial Requirements

During the workshop it was felt that accurate location of defects was one of the key criteria for successful condition assessment. The ability to inspect the pipeline, identify and quantify and assess the criticality of any defects, allowing the most appropriate remedial action to be taken at the most appropriate time, were the elements considered to be the most beneficial. For some applications, perhaps where company records were poor, it was also considered beneficial to be able to follow the line and level readings of any internal inspection device, as an aid to route mapping. However, for many utilities, especially in the US, it may not be one of the major criteria at this moment in time, as the use of systems such as GIS, are not yet fully utilized.

OPERATIONAL CONDITIONS

During the workshop, the options given by the utilities for the operation of any condition assessment device were (in order of preference) as follows:

- Utilize techniques that minimize any excavation by deployment through any existing access points such as valves or hydrants. These techniques should be non-interruptive; either utilized on the outside of the pipe or do not interrupt pressure and flow if inside the pipe.
- Utilize techniques that require a small excavation, but are still non-interruptive. This would include taking pipe coupons under pressure, the use of an external tool or technique to perhaps measure remaining pipe wall thickness and analysis of the backfill and pipe surround to identify any potential contamination or corrosive elements which, may lead to future failures.
- Utilize techniques that are interruptive, but may operate at reduced pressures within the pipeline, rather than having to isolate the main.
- Utilize techniques that require the main to be de-watered. This would include the traditional method of removing a pipe sample for laboratory analysis and also condition assessment techniques that require access to an empty main.

There are a number of considerations that need to be taken into account when considering the utilization of any condition assessment devices within the network. These operational considerations would include maintaining levels of service and control of supply, as well as the practical field implications of the use of these devices.

Impact on Levels of Service

Ideally, during any condition assessment of the pipe, the pipe should not need to be taken out of service. Acceptable levels of service must be maintained and minimum requirements are regulated in the U.K. and in the U.S.A. Water pressure and quality are of paramount importance to the customer, they rely on having water that is clean and wholesome and that is available in sufficient quantity on demand. Water companies would prefer to avoid any condition assessment method that would compromise water pressure and quality standards or take the pipe out of service. This is particularly important on trunk mains that supply water to large communities.

If internal inspections were necessary, then the tool must not affect water quality. There should only be minimal or controllable disturbance of any sediment or encrustation leading to potential discolored water problems and no foreign substances e.g. metals, plastics, lubricants etc. should be introduced into the water supply from the device that would exceed threshold levels set by regulators.

Network Control

Any potential condition assessment device should not adversely affect the control and operation of the network. Minimum operating pressures have to be maintained throughout the associated network. If it is necessary for a tool to enter the water pipe, it should not introduce a significant pressure drop from one end of the tool to the other. The tool should also be capable of traveling at the same velocity as the water moving through the pipe, to minimize any variations in flow, and allow any associated discharges to be controlled.

Field Operations

Ideally, in order to minimize any excavation, potential disruption and cost, any condition assessment tool should be able to access the pipe through any existing access points such as hydrants and valves. Any fittings necessary to aid access of the device to the pipe, should not reduce the integrity of the pipe and adhere to the specification of the pipe. The fittings must be able to meet the pressure requirements of the pipe i.e. it must be able to withstand the design pressure of the pipe. Any fitting must also comply with all safety standards (federal, state and in-house) and ideally the fittings should also be low cost.

The tool should also be able to operate successfully in water mains without the need for any excessive pre-cleaning operations to allow the tool to negotiate the pipe or to allow the tool to have intimate contact with the pipe wall. Some limited pre-cleaning may be possible using tools which can easily be inserted into the pipe, and any debris associated with the cleaning operation can be controlled. Easy recovery of the cleaning tool would also be desirable. Any inspection device, which requires the pipe to be aggressively cleaned or the pipe to be de-watered, would not be considered to be beneficial as the associated costs may prove to be prohibitive. Taking strategic or critical mains out of service proves to be very difficult and should be avoided.

For an in-pipe system, to obtain the most benefit from the inspection both in terms of economics and length of pipe covered, the tool should be able to negotiate bends, variations in pipe diameters and obstructions such as valves without the need to remove the potential obstruction. Again, the longer the length of pipe that can be surveyed per launch of the tool (perhaps in between two valves for example), the better the economics, benefit to the utility and the likelihood of these systems being used. It may be that some kind of simple pre-inspection tool is used to assess capability of the inspection tool to successfully negotiate the length of pipe under examination. Another important consideration is the time taken for the tool to complete the field inspection operation, especially if it has an impact on normal operations.

Table 4.2
Parameter values

Value	Definition
1	Not important
2	Desirable
3	Essential

In-Pipe Systems

Consideration was given in the workshop to the type and operation of any condition assessment devices. In-pipe systems were generally seen as a last resort. Water companies would rather not introduce something into the pipe that could potentially cause problems with the water supply or quality. The preference was to use technology that was completely non-intrusive requiring little or no excavation or direct access to the main. For example, infrared, which could be mounted onto a satellite or aircraft, or technology that can be used above ground by an operator, such as ground penetrating radar (GPR), or technology that can be mounted externally onto a pipe. In-pipe systems may be acceptable in certain applications, but if the pipe were a critical supply pipe then utilities would be more reluctant to use an in-pipe system in case of any problems and disruption to supply.

ASSESSMENT PARAMETERS

As mentioned previously, the assessment parameters are dependent on the failure characteristics of a particular pipe material. The parameters that are critical for say, ferrous pipes may only be desirable or even not important for PCCP pipes (Dickson 2003). For all of the different parameters considered during the workshop, they could generally be divided into four sets; the pipe itself; degradation mechanisms; other pipe aspects and operational parameters. Each parameter has been assigned a value of importance and these have been summarized in Table 4.2. Some value of accuracy has also been assigned to the parameters. This may be more of a "wish" list, as it may not be possible to achieve this level of accuracy either technically or economically.

One thing that was evident during the workshop was the need for better communication between the water utilities and the inspection tool providers to agree on what is required, what is possible to achieve and what is economically viable for the utility. There also needs to be a reduction in the levels of expectation driven by the providers and perhaps a better understanding from water utilities of what performance and results these tools can actually achieve. Equally as important is educating the research community about the unfilled science/technology gaps for water mains inspection and the associated benefits of addressing them. This may help to promote the interest of researchers and the support of research funding agencies for working on these topics.

The initial parameters considered aspects of the pipe itself and the possible information that could be obtained during the course of the inspection and assessment. Due to the relative importance of PCCP pipes at larger diameters, they have been considered separately to the other potential pipe materials. These pipe parameters are summarized in Table 4.3. Measurement of

Table 4.3
Pipe parameters

Parameter	Metallic/Polymer pipes value	PCCP pipes value	Accuracy
ID, OD	3	1	±5% WT
Wall thickness	3	1	±5% WT
Eccentricity	3	1	±5% WT
Ovality	3	1	±5% WT
Loss of wall section	3	1	±5% WT

Note: WT = wall thickness

these parameters would give an indication of the overall structural condition and capability of the pipeline to withstand the internal and external loading encountered during service, and would highlight any areas for concern.

Table 4.4 considers some of the potential degradation mechanisms for the different pipeline materials, which may give an indication of potential future failures. Corrosion pitting was highlighted as one of the most important factors for metallic pipes and a required level of accuracy was assigned to it.

A warning system that was able to analyze the depth of corrosion pits would be desirable, and could be set by the pipeline operator depending upon the circumstances. For example, if a pipe failed when the pipe wall was 80% corroded, then the tool should alert the operator when the pipe wall is 60-70% corroded so that an appropriate rehabilitation schedule could be drawn up (Dickson 2003) prior to any failure occurring. An assessment of the presence of any coating or lining and its condition would also help to highlight areas where potential degradation may occur in the future. Measurement of tuberculation as well as giving an indication of potential degradation of the pipe would also help to highlight areas where potential future hydraulic or water quality problems may occur. This was not seen as a major concern on larger diameter pipes but would be on pipes below 12 inch (300 mm). Leak detection was also considered to be an important parameter. It was felt that each utility would need to set the criteria to take into consideration what level of leakage was acceptable to them.

There were also other elements, which could provide useful information to the pipeline operator during any inspection process. Some of these are considered in Table 4.5.

Table 4.4
Degradation parameters

Parameter	Metallic/Polymer pipes value	PCCP pipe value	Accuracy
Corrosion pitting	3	3	±5% WT
Crack detection	3	3	Min crack length 20 mm ferrous and PVC
Leak detection	3	3	
Lining/coating detection	3	N/A	Determine breaks, disbondment
Tuberculation Assessment	1-2	1	Would be important for small diameter pipes

Table 4.5
Other pipe aspects parameters

Parameter	Value	Comments
Off-take detection	2	Difference between a corrosion hole and off-take
Material identification	3	
Pipe bedding info	2	Voids, hard spots
Feature location	3	±1m
Joint detection	3	Assessment of condition

It would also be beneficial if the tool were able to determine whether a perceived hole in a pipe was an off-take or whether the pipe had actually been completely corroded. There are also a number of other parameters beneficial for a water utility to obtain information on during any condition assessment process. Identification of the material type along the route of the pipeline under consideration would be useful for future reference as well as the potential to actually map the route and direction of the pipeline as the tool progresses. This may also help to identify old repairs, branches and connections (both legal and illegal) that may not be recorded. This information could then be used to accurately update the network data storage system (maps, plans, GIS etc.) operated by the utility. The status of the pipe bedding and the presence of any voids were important as the design of many pipe materials relies upon adequate compaction of the backfill to maintain full pipeline integrity. Also any areas where the backfill has subsided, may help to indicate potential future failure sites. Feature location will assist with maintaining accurate asset records, and help with the planning of any remedial work that may be required. Another important feature would be the ability to detect joints, and if possible, assess their sealing integrity.

There are a number of operational requirements, which need to be considered, some have already been discussed and these are summarized in Table 4.6.

The operational parameters identify the requirements of the utility to allow the device to successfully operate within the pipeline. Any in-pipe operating system would need to be able to withstand the pipeline design pressure (which may be higher than the actual operating pressure), and this capability may need to vary with pipeline diameter. The required operating temperature range would need to consider not only the range of water temperatures encountered, but also the environments around the pipe and above ground.

Table 4.6
Operational parameters

Parameter	Requirement
Access	Through existing features
Operating range	1 mile
Max operating pressure	Pipeline design pressure
Temperature range	30°F–110°F
Bend and bore passing	

REQUIREMENT SPECIFICATION

Needs Not Addressed

This section looks at findings from the study to ensure that the requirements for non-interruptive condition assessment inspection devices are met. Where there is still a technology deficit, or requirements are not met this is identified. Tables 4.7 through 4.9 compare assessment parameters against the various selected assessment technologies, providing a total score in each case. Separate tables are included for Metallic, Mineral based and Plastic pipe materials.

A high score indicates that essential requirements are being met. It should be noted that the tables make no attempt to weight individual condition assessment parameters in terms of their overall contribution. For different operating companies priorities will vary and change, it is important that the condition assessment inspection devices are able to respond to these changes. Instead, parameter values reflect the users "wish" to acquire information.

Table 4.7 demonstrates that there are a number of existing technologies capable of condition assessment inspection of metallic transmission mains. However, technologies such as CCTV, which report a high score, give us little information regarding the structural condition of pipes, other technologies such as BEM and RFEC cannot be considered as non-interruptive. The most viable technologies today are the external Long Range Ultrasonic and Magnetic Flux Leakage devices. Long Range Ultrasonic is limited to welded steel pipelines and has a limited inspection range of a few tens of feet (around 10 metres). Existing Magnetic Flux Leakage devices require an excavation to access the pipe, causing disruption and further cost.

Table 4.8 demonstrates that there are a number of existing technologies capable of condition assessment inspection of mineral-based transmission mains. In particular, PCCP transmission mains are well served. Remote Field Eddy Current/TC and Seismic Pulse Echo technologies can both be used to assess baseline condition but currently require the pipeline to be dewatered during inspection. Acoustic Emission technology is ideal for monitoring a pipeline for wire breaks and other acoustic events but requires one of the two technologies listed previously to determine baseline condition.

Table 4.9 demonstrates that there are only a very limited number of existing technologies capable of condition assessment inspection of plastic transmission mains. Assessment technologies are currently limited to Acoustic Leak Detection and CCTV inspection. There is currently no device available for the assessment of through-wall cracks, geometry or pipe bedding.

Table 4.7

Summary of assessment parameters versus technologies (metallic pipe)

Condition Assessment Parameters	Importance to Material Type	Acoustic Emission	Acoustic Leak Detection	Broadband Electromagnetic	CCTV	Long Range Ultrasonic	Magnetic Flux Leakage	Remote Field Eddy Current	Remote Field Eddy Current – TC	Seismic Pulse Echo	Ultrasonic
		Assessment Technology									
PIPE PARAMETERS											
ID, OD	3	-	-	-	-	-	-	-	-	-	-
As made wall thickness	3	-	-	3	-	-	3	3	-	-	3
Eccentricity	3	-	-	3	-	-	3	-	-	-	3
Ovality	3	-	-	-	-	-	-	-	-	-	-
DEGRADATION PARAMETERS											
Remaining wall thickness	3	-	-	-	-	3	3	3	-	-	3
Corrosion pitting	3	-	-	3	-	-	3	3	-	-	3
Crack	3	-	-	-	-	-	-	-	-	-	-
Leak	3	-	3	-	-	-	-	-	-	-	-
Loss of lining/coating	3	-	-	-	3	-	-	-	-	-	-
Tuberculation assessment	1	-	-	-	3	-	-	-	-	-	-
Delamination	3	-	-	-	-	-	-	-	-	-	-
Wire breaks	3	-	-	-	-	-	-	-	-	-	-
Chemical degradation	3	-	-	-	-	-	-	-	-	-	-
FEATURE DETECTION											
Off-take detection	2	-	-	-	2	-	-	-	-	-	-
Material identification	3	-	-	-	-	-	-	-	-	-	-
Pipe bedding info	2	-	-	-	-	-	-	-	-	2	-
Feature location	3	-	3	-	3	-	-	-	-	-	-
Joint detection	3	-	-	-	3	3	-	3	-	-	-
TOTAL SCORE		-	6	9	14	6	12	12	-	2	12

Note: the – means not applicable

Table 4.8

Summary of assessment parameters versus technologies (mineral based pipe)

Condition Assessment Parameters	Importance to Material Type	Acoustic Emission	Acoustic Leak Detection	Broadband Electromagnetic	CCTV	Long Range Ultrasonic	Magnetic Flux Leakage	Remote Field Eddy Current	Remote Field Eddy Current/TC	Seismic Pulse Echo	Ultrasonic
	Assessment Technology										
PIPE PARAMETERS											
ID, OD	3	-	-	-	-	-	-	-	-	-	-
Wall thickness	1	-	-	-	-	-	-	-	-	-	-
Eccentricity	1	-	-	-	-	-	-	-	-	-	-
Ovality	1	-	-	-	-	-	-	-	-	-	-
DEGRADATION PARAMETERS											
Remaining wall thickness	1	-	-	-	-	-	-	-	-	-	-
Corrosion pitting	3	-	-	-	-	-	-	-	-	-	-
Crack	3	-	-	-	-	-	-	-	-	-	-
Leak	3	-	3	-	-	-	-	-	-	-	-
Loss of lining/coating	-	-	-	-	-	-	-	-	-	-	-
Tuberculation assessment	1	-	-	-	1	-	-	-	-	-	-
Delamination	3	-	-	-	-	-	-	-	-	3	-
Wire breaks	3	3	-	-	-	-	-	-	3	3	-
Chemical degradation	3	-	-	-	-	-	-	-	-	3	-
FEATURE DETECTION											
Off-take detection	2	-	-	-	2	-	-	-	-	-	-
Material identification	3	-	-	-	-	-	-	-	-	-	-
Pipe bedding info	2	-	-	-	-	-	-	-	-	2	-
Feature location	3	-	-	-	3	-	-	-	-	-	-
Joint detection	3	-	-	-	3	3	-	3	-	-	-
TOTAL SCORE		3	3	-	9	3	-	3	3	11	-

Note: the – means not applicable

Table 4.9

Summary of assessment parameters versus technologies (plastic pipe)

Condition Assessment Parameters	Importance to Material Type	Assessment Technology: Acoustic Emission	Acoustic Leak Detection	Broadband Electromagnetic	CCTV	Long Range Ultrasonic	Magnetic Flux Leakage	Remote Field Eddy Current	Remote Field Eddy Current/TC	Seismic Pulse Echo	Ultrasonic
PIPE PARAMETERS											
ID, OD	3	-	-	-	-	-	-	-	-	-	-
Wall thickness	3	-	-	-	-	-	-	-	-	-	-
Eccentricity	3	-	-	-	-	-	-	-	-	-	-
Ovality	3	-	-	-	-	-	-	-	-	-	-
DEGRADATION PARAMETERS											
Remaining wall thickness											
Corrosion pitting	3	-	-	-	-	-	-	-	-	-	-
Crack	3	-	-	-	-	-	-	-	-	-	-
Leak	3	-	3	-	-	-	-	-	-	-	-
Loss of lining/coating	3	-	-	-	-	-	-	-	-	-	-
Tuberculation assessment	1	-	-	-	-	-	-	-	-	-	-
Delamination	3	-	-	-	-	-	-	-	-	-	-
Wire breaks	3	-	-	-	-	-	-	-	-	-	-
Chemical degradation	3	-	-	-	-	-	-	-	-	-	-
FEATURE DETECTION											
Off-take detection	2	-	-	-	2	-	-	-	-	-	-
Material identification	3	-	-	-	-	-	-	-	-	-	-
Pipe bedding info	2	-	-	-	-	-	-	-	-	-	-
Feature location	3	-	-	-	3	-	-	-	-	-	-
Joint detection	3	-	-	-	3	-	-	-	-	-	-
TOTAL SCORE			3		8						

Note: the – means not applicable

OUTLINE SPECIFICATION

This outline specification should act as a design source to manufacturers and suppliers of non-interruptive condition assessment inspection devices. Performance has been specified in accordance with requirements developed as part of the project workshop. It is recognized that the list is not exhaustive, but reflects those parameters considered most important.

Device Operational Features

Access to Pipe

For in-pipe systems, the device will be deployed into the main through existing fittings.

Operating Range

The device shall be capable of 1-mile (1.6-km) range within one working day.

Operating Pressures

Maximum operating pressure of the device shall be in accordance with the pipeline design pressure.

Operating Temperature Range

The device shall be capable of operating to specification over a temperature range of 30°F to 110°F (-1.0 °C to 43°C).

Bore Passing Capability

The device shall be capable of passing a uniform diameter restriction of 15% of nominal bore.

Bend Passing

The device shall be capable of passing 90-degree bends with a centerline radius of 1× pipe diameter.

Pipe Parameter Measurement

Pipe Diameter

Pipe internal diameter shall be estimated to within ±5% of nominal wall thickness. Pipe external diameter shall be estimated to within ±5% of nominal wall thickness.

As-Made Wall Thickness

This is an estimate of the wall thickness of the pipe immediately post construction. The made wall thickness shall be estimated to within ±5% of nominal wall thickness.

Bore Eccentricity

Bore eccentricity shall be estimated to within ±5% of nominal wall thickness.

Ovality

Ovality shall be estimated to within ±5% of nominal wall thickness.

Degradation Parameter Assessment

Corrosion Depth Evaluation

Accuracy of corrosion depth estimation shall be ±5% of nominal wall thickness.

Crack Detection and Measurement

The device shall be capable of detecting longitudinal and circumferential cracks, in ferrous and PVC pipes, greater than ¾ inch (20 mm) in length.

Leak Detection

Leaks shall be detected

Pipe Lining

The device will be capable of detecting cracks and disbondment in internal linings and coatings

Tuberculation

The device will identify tuberculation.

Other Parameters

Feature Detection

The device will identify pipe features:

Off-takes (and distinguish them from holes).
Joints

Material Identification

The device will identify and differentiate different pipe materials

Pipe Bedding

The device will identify voids and hard-spots.

Location Accuracy

With respect to location of features, these shall be reported to within ±3.28-ft (±1 m).

CHAPTER 5
CONCLUSIONS AND RECOMMENDATIONS

CONCLUSIONS

- The technology review undertaken within this project was able to build upon previous work carried out by AwwaRF. The outputs presented are, however, more comprehensive than previous studies as they identify additional available technologies being utilized and new products and services under development.
- The study highlighted the importance of understanding the business needs of water utilities and provides an overview of some of the key business drivers and justifications.
- It is important to understand how the available technologies can assist in the decision-making process for water utilities. The application of these tools and the results obtained can contribute to the asset management process. However, there are varying levels of application and strategies existing.
- Although the commodities and the business drivers are generally different within the oil and gas industries, there are some links with the water industry. This is currently particularly applicable in the area of risk management of the steel assets.
- One of the strengths of this project was the ability to consider the subject on a global basis.
- The workshop provided an ideal forum to expand the understanding of the business drivers for the Water Industry and the requirements of the utilities for condition assessment. The strong panel of experts assembled from water utilities, technology providers, consultants, researchers, academia, regulators and government departments were also able to review the applications of existing technologies and contribute to the ideal parameters for a non-interruptive condition assessment inspection device.
- A number of traditional technologies are being successfully used by some water utilities to assist them with their decision making process:
    - Pipeline history
    - Location of mains
    - Pressure testing
    - Flow tests and Hazen Williams C tests
    - Soil resistivity
    - Coupons and samples
- The experts attending the workshop contributed to the development of a generic parameter portfolio for an ideal specification. No single device exists to meet the requirements of this generic specification or all of the existing materials in use.
- A number of the existing technologies go some way to address some of the generic requirements:
    - Acoustic Emission Monitoring
    - Acoustic Leak Detection
    - Broadband Electromagnetic
    - CCTV
    - Long Range Ultrasonic (Guided Wave)

- Magnetic Flux Leakage
- Remote Field Eddy Current
- Remote Field Eddy Current – Transformer Coupling
- Seismic Pulse Echo
- Ultrasonic

Not all of the above technologies are truly non-interruptive condition assessment inspection devices.

- It was not possible within the scope of this project to rank the potential varying assessment parameters.
- The adoption of these technologies within the Water Industry seems to be limited. Although in certain cases, the PCCP assessment technologies appear to be providing solutions for customers. Some elements which may be affecting the adoption of the technologies on a more widespread basis could be:
    - Charging arrangements for service
    - Performance of technologies
    - Expectation/capability of results

RECOMMENDATIONS

- Water utilities clearly need to define their business needs and adopt strategies to meet their individual operational circumstances. A good understanding of data management needs is essential.
- Technology providers need to review their methods of charging for services to closer suit the business needs of water utilities
- No single non-interruptive condition assessment inspection device exists for all water transmission mains. Water utilities should consider each job on its merits and select from the most appropriate technology available to provide the required results:
    - Traditional technologies
    - Existing condition assessment technologies
- It is recommended that further research is undertaken to take full advantage of the outputs from this project, and facilitate the transfer of knowledge in this important area of work:
    - Additional consultation would be desirable as follows:
        1. To expand the understanding of the needs and practices of water utilities
        2. To review further the customers' experience from the application of existing technologies (close collaboration would be required with the technology providers)
        3. To maintain links with the technology providers on service enhancement and new applications in the field
        4. To explore further potential research developments with the academia
    - Enhance and maintain the project web site as a condition assessment forum:
        1. Case studies
        2. Technology providers links

3. Events
4. Discussion forum

- Facilitate on-going workshops dedicated to condition assessment:
    1. Event every 2 years

    Water Utilities/Technology Providers/Consultants

    Outputs to be made available by AwwaRF on CD-ROM
- Develop a detailed specification for a non-interruptive condition assessment inspection device by material:
    1. Further consultation with the Water Utilities to define ranking criteria for the identified parameters
    2. Liaison with the Technology Providers to assess the suitability of existing technologies to meet the developed specification and understand development opportunities
- Develop a Best Practice Selection Manual for the Condition Assessment of Water Mains:
    1. Review and scope of existing technologies
    2. Selection criteria of technologies for different materials
    3. List of service providers

APPENDIX A
COMPANIES CONSULTED

The authors of this report would like to thank all the utilities for taking the time to participate in the telephone consultation. These companies are listed below:

American Water
Blacksburg, Christiansburg, VPI Water Authority
Boston Water and Sewer Commission
Central Arkansas Water
City of Chicago Department of Water
City of Cleveland – Division of Water
City of Dayton
City of Halifax Water Commission
City of Houston Public Utilities Division
City of Lincoln Public Utilities Water Division
City of Manchester Water Works
City of Ottawa
City of Phoenix
City of Tampa Water Department
City of Virginia Beach
Dallas Water Utilities
Denver Water
Fairfax County Water Authority
Grand Strand Water & Sewer Authority
Greater Cincinnati Water Works
Jefferson Utilities
Kansas City Board of Public Utilities
Los Angeles Department of Water and Power
Massachusetts Water Resource Authority
Metropolitan Water District of Southern California
Monroe County Water Authority
New York State Public Service Commission
Northumbrian Water Limited
Orange County Utilities Water Division
Philadelphia Suburban Corporation, Inc.
Philadelphia Water Department
Portland Water District
Providence Water
San Francisco Public Utilities Commission
South Central Connecticut Regional Water Authority
Tallahassee Water Utility
Thames Water
United Utilities
United Water

Washington Suburban Sanitary Commission
Yorkshire Water Services

APPENDIX B
OTHER COMPANIES INVITED TO PARTICPATE IN THE CONSULTATION

Albuquerque Public Works
Anne Arundel County Department of Public Works
Board of Water Supply
Calleguas Municipal Water District
Charleston Commissioners of Public Works
City of Austin Water and Wastewater
City of Houston
City of Jacksonville Water Supply
City of Oklahoma City
City of San Diego
City of Toledo Water Division
Cobb County Water System
Cobb County-Marietta Water Authority
Colorado Spring Utilities
DeKalb County Public Works
Des Moines Water Works
East Bay Municipal Utilities District
El Paso Water Utility
Elizabethtown Water Company
Erie County Water Authority
Evaston Water Department
Hartford Metropolitan District
Illinois-American Water Company
Indianapolis Water Company
Jefferson Parish Water Department
Kansas City Water Services Department
Las Vegas Valley Water District
Long Beach Water Department
Louisville Water Company
Memphis Light Gas & Water
Metropolitan Water District
Metropolitan Water District of Southern California
Miami-Dade County Water and Sewer
Milwaukee Water Works
Newark Division of Water and Waste
Omaha Metropolitan Water District
Philadelphia Suburban Water Corporation
Regional Water Authority
Richmond Department of Public Utilities
Salt Lake City Public Utilities
Salt Lake County Water Conservancy

San Diego County Water Authority
San Jose Water Company
Seattle Water Department
St. Paul Water Utility
Suffolk County Water Authority
Tarrant County Water District
Tucson Water
Tulsa Public Works Department

APPENDIX C
UTILITIES INVITED TO PARTICIPATE IN THE WORKSHOP

Aguas Andinas
Central Arkansas Water
Charleston Community of Public Works
City of Chicago
City of Ottawa
City and County of San Francisco
Elizabethtown Water Company
Greater Cincinnati Water
Los Angeles Department of Water and Power
Lyonnaise des Eaux France
Northumbrian Water Limited
Philadelphia Suburban Water Company
Severn Trent Water
Thames Water
United Water
Washington Suburban Sanitary Commission
Welsh Water
Yorkshire Water Services

APPENDIX D
LIST OF DELEGATES

Table D.1
List of delegates who attended the workshop

Name	Company	Country	Role
Lillie, Kevin	NWTC	U.K.	Project Team
Daniels, Simon	Issquared	U.K.	Project Team
Smart, David	Issquared	U.K.	Project Team
Reed, Chris	NWTC	U.K.	Project Team
Cowan, Rob	City of Ottawa	Canada	Project Partner
Morrow, Michael	NWL	U.K.	Project Partner
Kopp, Gunnar	GE Power Systems	U.K.	Project Partner
Zhang, Jian	AwwaRF	U.S.A.	Client
Proctor, John	Yorkshire Water Services	U.K.	PAC Member
Wtorkowski, Albert	City of Chicago	U.S.A.	PAC Member
Royer, Mike	USEPA	U.S.A.	PAC Member
Russom, Dale	Central Arkansas Water	U.S.A.	Water Industry
Bush, Cheri	Greater Cincinnati Water Works	U.S.A.	Water Industry
Gomez, Enrique	Los Angeles Department of Water and Power	U.S.A.	Water Industry
Attia, Ramsis	City and County of San Francisco	U.S.A.	Water Industry
Gaw, Cary	United Water	U.S.A.	Project Partner
Glozzy, Jim	United Water	U.S.A.	Project Partner
Hartman, Bill	Hydroscope Inc., USA	U.S.A.	Technology Provider
Dickson, Dudley	AESL	U.K.	Technology Provider
Wayman, Malcolm	AESL	U.K.	Technology Provider
Mergelas, Brian	PPIC Ltd.	Canada	Technology Provider
Henrich, Gord	PPIC Ltd.	Canada	Technology Provider
Holley, Mark	Pure Technologies U.S. Inc.	U.S.A.	Technology Provider
Elliott, Jack	Pure Technologies Ltd.	Canada	Technology Provider
Thomson, James	American Investigation and Assessment Inc.	U.S.A.	Technology Provider
Hazelton, Glyn	American Investigation and Assessment Inc.	U.S.A.	Technology Provider
Meegoda, Jay	New Jersey Institute of Technology, Newark	U.S.	Research Industry

(continued)

Table D.1 (Continued)

Name	Company	Country	Role
Juliano, Tom	New Jersey Institute of Technology, Newark	U.S.A.	Research Industry
Woodcock, Mike	Washington Suburban Sanitary Commission	U.S.A.	Water Industry
Harstead, Gary	United Water	U.S.A.	Water Industry
Neilan, Thomas	United Water	U.S.A.	Water Industry
Woodward, Tony	Thames Water	U.K.	Water Industry
Anderson, Rodney	National Energy Technology Laboratory, US DOE	U.S.A.	Research Industry
Micklus, Drew	Freyssinet	U.S.A.	Technology Provider
Stubler, Jerome	Freyssinet International	France	Technology Provider
Tessier, Marc	Freyssinet	U.S.A.	Technology Provider
Rajani, Balvantrai	National Research Council Canada	Canada	Research Industry
Delescinskis, T.	United Water	U.S.A.	Water Industry
Winn, Kevin	United Water	U.S.A.	Water Industry
Patel, Manoj	Elizabethtown Water Company	U.S.A.	Water Industry
Biehler, Bob	Elizabethtown Water Company	U.S.A.	Water Industry
Krupa, C	United Water	U.S.A.	Water Industry
Radice, Pat	United Water	U.S.A.	Water Industry
Schink, Thomas	United Water	U.S.A.	Water Industry
Brandt, Stanislas	Advitam	U.S.A.	Technology Provider
Von Autenried, Ronald	Buck, Seifert & Jost, Inc.	U.S.A.	Consultant
Oberi, Kanwal	Charleston Community of Public Works	U.S.A.	Water Industry
Khiari, Djanette	AwwaRF	U.S.A.	Client
Bubenko, J	United Water	U.S.A.	Water Industry

APPENDIX E
WATER UTILITIES' FEEDBACK

Table E.1
Logistics

	Description	Score
1	Pre-Notification of objectives	3.85
2	Right mix of attendees	4.42
3	Quality of advanced information	3.57
4	Quality of workshop pack	4.28
5	Venue and facilities	4.28
6	Hospitality	5.00
	Average	4.23

Table E.2
Structure and content of programme

	Description	Score
7	Quality of relevance of presentations	4.14
8	Support of poster sessions	3.50
9	Success of work group forums	3.85
10	Opportunities for networking	4.57
11	Awareness of next steps	3.85
12	Workshop objectives achieved	3.85
	Average	3.96

Table E.3
Conclusions

	Description	Score
13	Benefit of workshop to you	4.42
14	Benefit of workshop to your company	4.42
	Average	4.42
	Overall Average	4.20

APPENDIX F
TECHNOLOGY PROVIDERS' FEEDBACK

Table F.1
Logistics

	Description	Score
1	Pre-Notification of objectives	3.0
2	Right mix of attendees	3.0
3	Quality of advanced information	2.6
4	Quality of workshop pack	3.3
5	Venue and facilities	3.6
6	Hospitality	4.6
	Average	3.35

Table F.2
Structure and Content of Programme

	Description	Score
7	Quality of relevance of presentations	2.6
8	Support of poster sessions	2.0
9	Success of work group forums	2.6
10	Opportunities for networking	3.3
11	Awareness of next steps	3.0
12	Workshop objectives achieved	2.0
	Average	2.58

Table F.3
Conclusions

	Description	Score
13	Benefit of workshop to you	3.0
14	Benefit of workshop to your company	2.3
	Average	2.65
	Overall Average	2.86

APPENDIX G
CONSULTANTS & ACADEMIA FEEDBACK

Table G.1
Logistics

	Description	Score
1	Pre-Notification of objectives	4.3
2	Right mix of attendees	4.3
3	Quality of advanced information	4.6
4	Quality of workshop pack	5.0
5	Venue and facilities	4.6
6	Hospitality	5.0
	Average	4.63

Table G.2
Structure and Content of Programme

	Description	Score
7	Quality of relevance of presentations	4.3
8	Support of poster sessions	3.3
9	Success of work group forums	4.0
10	Opportunities for networking	4.3
11	Awareness of next steps	4.3
12	Workshop objectives achieved	4.3
	Average	4.08

Table G.3
Conclusions

	Description	Score
13	Benefit of workshop to you	4.0
14	Benefit of workshop to your company	3.3
	Average	3.65
	Overall Average	4.12

APPENDIX H
CASE STUDY – WATER UTILITY

CASE STUDY: NORTHUMBRIAN WATER

Strategic Mains Programme Strategy

The strategy for the Strategic Mains Program commences with a topographical survey of the feeder main to the Schedule 19 targeted District Meter Areas (DMA's). The bulk of this work, as well as that carried out on site, constitutes compiling and searching out all previous data and drawings relating to the pipeline in question. After the site survey has been completed and a definitive plan drawn up then the Company's GIS is updated accordingly. The site survey identifies, among others, areas of work to be carried out under other parts of the programme – maintenance issues and strategic crossings. The intention is that these other projects then identify further work to be undertaken. Under the maintenance heading this tends to be chamber re-builds. From the Strategic Crossings, it is expected to be necessary to carry out some fairly substantial mainlaying/duplication schemes. These are to be known as Named Projects. Also under this banner will be a sub program for pipeline coatings and repairs which become evident through the various site visits/surveys, as well as analysis points. The latter are to be strategically placed along the main to provide access for equipment, such as turbidity monitors, sampling, quadrinas (insertion meters) etc.

The main objective of the program is to survey and make records of and establish the location of valves, chambers and fittings on all strategic mains, as well as their location and structural integrity where they cross rail lines, roads and rivers.

Regional Strategic Crossings Sub Program

Objectives

To determine and understand the level of risk posed at water mains crossings in a systematic and controlled way. The key objective in this program is to manage the level of risk of loss of supply to our customers.

Scope

The scope of the program covers all strategic potable water mains that are perceived to be vulnerable, i.e. where a single supply main crosses over or under either a river, trunk road or a railway line, or any other vulnerable location.

Exposed Pipes

In the case of exposed pipes the survey includes a general assessment of the crossing including its overall condition. This includes notes on the degree of visible external metal loss and pit size. The ground/pipe interface and condition are described along with the number and type of supports or hangers and notes on the restraint provided.

A description of the number and type of joints is made. Where these are exposed an assessment of their condition and any deflection from their original alignment is made.

Information from topographical and network surveys is also included in the initial site survey where available, together with a rating of the crossing's priority.

Buried Pipes

A buried pipeline route will involve various load-bearing or unstable ground conditions, examples including marshy areas, subsidence or mining activity. The pipeline must accommodate any ground movement and not attempt to resist it. A protective coating of either a bitumen wrap for steel or factory-applied bitumen for iron piping is suitable for the majority of soil conditions.

Buried mains at strategic crossings are often protected using a sleeve or casing around the pipe itself and this is described from details contained in NWL's as-built drawings (if available). An inspection of the visible casing is made, if exposed by the excavations. If a cathodic protection system is shown on the drawings this is noted in the report.

Clearly an inspection of the joints is not possible in the case of buried mains unless the inspection excavations are located at these positions. Similarly, a full geological survey to assess the risk of ground movement is beyond the scope of these procedures.

Non-Destructive Testing

The wall thickness measurement is undertaken using a standard non-destructive testing (NDT) technique, which causes no damage or intrusion into the pipe.

Magnetic Flux Leakage

This is carried out using the AES Ltd., External Condition Assessment Tool (ECAT), which allows rapid non-intrusive assessment of the pipe wall thickness of iron and steel pipelines. The system is designed for operation in pipe sizes up to 49-inch (1,250-mm) diameter.

The device is based on conventional magnetic flux leakage technology. In this case a magnetic field is produced in the pipe wall using coils attached to the unit. The field is proportional to the wall thickness and defect size and is detected using sensors built into the tool, which convert the field into a voltage. These voltages are then converted to digital data, which is captured and stored using the system's on-board electronics and solid-state memory. The data is related to actual wall thickness via calibration curves.

At the end of an inspection on a particular crossing the data is downloaded to a computer. The information is then displayed numerically or by contour plots, as required.

The ECAT takes continuous readings at each point as it is traversed along the pipe. Multiple inspection sweeps are made around the circumference using a fixed-width shoe, until a representative sample is achieved.

Lord McGowan's Bridge

Conduit C62 is a 24-inch (600-mm) steel pipe and carries potable water from Lackenby Reservoir at 50.3 m to Redcar via Lord McGowan's Bridge on the A1085 at 8.6 m, giving a static head of 41.7 m. It was built circa 1965. The crossing is considered a top priority, being a strategic main crossing chemical pipework, a road and a railway. Access to the ground interfaces is down a steep bank from the A1085. Access to the crossing requires Enron and Corus permission, with appropriate Health and Safety arrangements as necessary.

The visual condition of the main gave sufficient cause for concern for it to be assessed, being the sole potable water supply for approximately 100,000 customers.

On the water main itself, two smoke deflector plates were attached to offer protection against diesel exhaust fumes from the locomotives that pass beneath the conduit. One of these smoke plates corroded sufficiently to become partly detached and hang from the water main over the rail line.

ICI brought in a "cherry picker" (mobile lifting platform), to obtain access to remove the partly fallen smoke deflector plate and inspect the second smoke deflector plate.

AESL were asked by NWL to carry out a provisional inspection of the conduit in order to determine the condition of the pipe crossing, and recommend any remedial actions necessary.

The pipe is exposed at this crossing, almost completely uncoated and badly corroded. It has a minimum measured wall thickness of 5 mm and the deepest external pit found was 5.1 mm deep, suggesting the imminent possibility of leaks. Before and during remedial work undertaken in 2001, several small leaks occurred.

Conclusions From the Condition Assessment

- The remaining smoke deflector was at risk of becoming detached from the water main.
- The entire protective coating had failed along the exposed conduit.
- General corrosion had taken place along the entire exposed section of the conduit.
- The conduit had suffered moderate to severe corrosion at the pipe ground interface.

The Lord McGowan's Bridge scheme (Figure H.1) typifies the condition assessment of strategic mains sub-programme known as Strategic Crossings. This water main is 24-inch steel and crosses ICI process pipework, two rail lines and a private road, above ground, supported on dedicated concrete piers. The main is the sole feed into the Redcar area supplying some 100,000 customers. The condition assessment demonstrated that the main was structurally unsound, had a minimum lifespan and thus was unfit for purpose. As a result the main was successfully replaced and duplicated as a "Named Project" to the sum of £350,000 (Figure H.2).

Figure H.1 Lord McGowan's Bridge – the original crossing (*Source*: Morrow, M., *Non-Intrusive Condition Assessment of Water Pipelines, 2003*)

Figure H.2 The successfully replaced duplicated mains (*Source*: Morrow, M., *Non-Intrusive Condition Assessment of Water Pipelines, 2003*).

APPENDIX I
CASE STUDIES

CASE STUDY 1: NON-DESTRUCTIVE TESTING OF WATER MAIN SUPPLY LINES

Manufacturer/Patent holder: Rock Solid Pty. Ltd.
Assessment technology: BEM

As part of its asset condition evaluation program City West Water Ltd. engaged Rock Solid (RSG) to undertake non-destructive testing of water distribution mains throughout Melbourne. The mains consisted of mild steel, ductile iron; cement lined cast iron and plastic lines. Some of the lines have been in use for nearly a century raising concerns about their condition.

Repairs and replacement of water distribution infrastructure for most water authorities consists of spot repairs of isolated bursts, lining or re-lining of existing pipes as part of the infrastructure upgrading or complete replacement when frequency of failures is high.

In order to obtain a better understanding of the assets' conditions to enable planning and targeting of upgrades with limited budgets, non-destructive testing of cement lined cast iron (CLCI) pipes was undertaken to assess the pipe wall's thickness, graphitisation and general pipe condition.

Based on past experience two techniques developed by RS were used to cross correlate data for a more accurate survey result. Broadband Electromagnetics (BEM) and Rapid Magnetic Permeability Scanning (RMPS) were used to determine the overall conditions of the pipes with specific emphasis on metal loss and graphitisation.

The CLCI pipes were opened and the probes pulled through the pipes via a cable that had been rodded or floated through the pipe length. Measurements are continuously taken while the instrumentation travels through the pipe slowly obtaining measurements in 4 inch (100 mm) intervals.

The non-destructive survey data is now being collected together with geological, geotechnical and metallurgical developing a databank for future infrastructure budgeting and planning as well as developing a useful tool for a comprehensive audit of the water supply network.

CASE STUDY 2: A MANAGED APPROACH TO PRESTRESSED CONCRETE CYLINDER PIPE

Manufacturer/Patent holder: Washington Suburban Sanitary Commission
Assessment technology: Seismic Pulse Echo

A managed approach to the problem of deteriorating prestressed concrete cylinder pipe (PCCP) has been applied to over 27,500 feet (8,382 m) of a 96-inch (2.4-m) diameter water transmission main. The evaluation of the pipeline was conducted in four phases; review of historical information on design and construction, internal visual inspection and manual sounding of the pipe, nondestructive testing (NDT), and a combined analysis and assessment of the accumulated information. The results of the water main evaluation conducted provided a logical and repeatable tool to assess the various options for operation of the pipeline. The tool considers customer demand, existing capital investment, costs associated with pipeline rehabilitation or replacement, the physical condition of the line, and the relative risks to the public. Using the pipeline evaluation methodology presented, a decision was taken not to reline the entire 96-inch (2.4-m) water with welded steel as originally planned. The rehabilitation approach adopted is estimated to result in capital savings in excess of 30 million dollars.

Introduction

Water and sewer utilities are increasingly facing the problems associated with infrastructure deterioration. Of particular concern are underground pressure mains, both because of the inherent difficulties associated with inspecting a buried structure and the fact that their alignments often extend through heavily developed and populated areas. In the case of large diameter pipelines, considerable capital has been invested which should be utilized as efficiently as possible without incurring undue costs and risk. The composite structure of PCCP further complicates any evaluation program and requires an approach that combines the best historically proven inspection techniques with the evolving methodologies of NDT. An effective management approach has been applied to the investigation of a 96-inch (2.4-m) diameter PCCP water transmission main. This approach considers the need to meet existing and anticipated customer demands, optimize additional capital expenses associated with pipeline operation, and minimize the risk to both the public and their property from the possibility of a rupture in the main.

Configuration and Construction of Prestressed Concrete Cylinder Pipe

PCCP is a composite product that relies on the interaction of the various construction materials to provide satisfactory performance. A thin gage steel cylinder is constructed, using either spiral or longitudinally welded seams. Steel bell and spigot rings for a rubber gasketed joint are then welded to the ends of the steel cylinder. The steel cylinder and rubber gasketed joint rings serve the purpose of making the assembled pipeline watertight. In the case of embedded cylinder type PCCP, of which the subject 96-inch (2.4-m) pipeline is constructed, inner and outer concrete cores are then cast around the steel cylinder. After the concrete core is cured, a high strength prestress wire is wrapped around the pipe at a design-specified tension and spacing, forming a continuous spiral of wire. The strength of the PCCP is dependent on compression of the concrete core and steel cylinder produced by the prestressing operation. After the prestressing operation, a Portland cement mortar mix is applied over the prestress wire for corrosion protection through encasement in a highly alkaline environment. The mortar coating is typically specified as 13/l6-inch thick, but on the various contract that comprise the 96-inch pipeline varies from 13/l6 inch (22 mm) to 1.5 inches (38 mm.) in thickness. There were no provisions made in the design of this pipeline to bond joints between individual pipe lengths or to provide electrical continuity between the prestress wire and the steel cylinder through the use of shorting straps.

Pipeline Evaluation

Review of Pipeline History

The initial phase of the PCCP transmission main investigation entailed a review of existing information on the pipeline. Plans, specifications, construction records, manufacturer's submittals, and other documentation pertaining to the line were gathered and reviewed. This provided an intimate knowledge of job specific pipeline details that proved invaluable in the interpretation of anomalous pipe conditions encountered during the later stages of the investigation. Since PCCP is designed for the particular project, the pipe manufacturer's pipe design specifications and pipe laying schedules were extremely important in that they were the most accurate record of the actual pipe construction and installation details. Construction correspondence was reviewed which contained information on installation methodology, problems encountered on the project, and deviations from the original design. Data on the pipe construction materials, consisting of both specifications from the manufacturer and test results from other investigations, were beneficial in that many of the historical problems associated with PCCP are related to the pipe construction materials.

Internal Visual Inspection and Manual Sounding of the Pipe

Openaka Corporation, Inc. conducted an internal inspection, where any visible signs of distress or other anomalies were documented. During the inspection each individual piece of pipe was uniquely numbered to simplify record keeping. The inner concrete core was sounded for hollows that are often indicative of broken prestress wire. Visible cracks in the inner concrete core exemplify the need for an experienced inspector because some types, such as circumferential prestressing cracks and spigot cracks, are common in this type of pipe and are normally not an indicator of problems. Longitudinal cracks in certain locations, however, are often reflections of broken prestress wire and pipe distress. Similarly, hollows in the inner concrete core indicate a loss of compression that can be the result of prestress wire failure. Past experience internally inspecting other pipelines at the Washington Suburban Sanitary Commission (WSSC) and other utilities in the last decade has shown that these techniques of internal inspection can detect deteriorated pipe that is at risk of catastrophic failure.

Non-destructive Testing of the Pipe

In order to detect possible pipe distress prior to when techniques employing visual inspection and manual sounding are effective, non-destructive testing of the 96-inch (2.4 m) PCCP water main was conducted from inside the pipeline. This testing utilized sonic/ultrasonic techniques and equipment that are being cooperatively developed by the WSSC and NDT Engineering, Inc. Analyses of the data have shown that the NDT can detect anomalous pipe characteristics, such as those produced by broken prestress wires, prior to when they are detectable using more conventional methods. Conversely, the NDT can be used to identify pipe lengths that are in good condition. Additional work will be required before unique characterization of the NDT signatures of various defects and their relation to the pipeline rate of decay are established. The NDT results to date have been promising in that even though the signal characteristics representative of structural distress cannot be definitively distinguished from anomalies that may be the result of manufacturing or installation damage. However they still allow each individual pipe length to be classified as pipe that is good or pipe that is potentially bad. This in itself is an improvement over other PCCP evaluation techniques that extrapolate pipeline condition from a limited number of tests conducted on a small sample population.

Data Reduction and Analysis

Correlation of Pipeline Data with Inspection Results

A computer spreadsheet was set up to correlate the unique pipe numbers established for the investigation with the section's location along the pipeline, any known defects, inspection observations, and the NDT information. A description of each length of pipe and all fittings was entered into the spreadsheet. Other data included the unique pipe number assigned during the investigation, the beginning and ending pipe stations, class of PCCP, inspection observations, and any other pertinent physical features such as concrete encased sections or pipe located within tunnels. This served as the foundation of the pipeline evaluation and all of the data analyzed was eventually entered into the spreadsheet. This approach proved to be beneficial in the overall condition assessment, and especially in the interpretation of the NDT data, since all of the information was consolidated in one place for each of the 1,434 individual pipe sections within the line.

Repair Estimates for Correcting Known Pipeline Deficiencies and Defects

The review of historical information on the 96-inch (2.4 m) PCCP water main, which included previous inspection reports and engineering analyses, was combined with the results of the current internal visual inspection and manual sounding of the pipeline. This produced a list of deficiencies and defects that required corrective action. Planning level cost estimates were then made for each of the repair items as well as the cost for relining or replacing the lengths judged to be defective. The repair estimates were then divided into costs that would be incurred only if the individual pipe was to remain in service and those that would still remain if the pipe was structurally relined or replaced. A further subdivision of repair costs was made to separate out certain pipeline features that in past design evaluations have been classified as marginal but still serviceable under the pipeline operating conditions.

Evaluation of NDT Data

Each NDT record collected consists of four individual sensor traces accumulated using an array measuring 2.5 feet (0.76 m) in length. Over 88,000 individual NDT traces were collected and evaluated in the investigation. The traces were collected in two continuous lines on either side of the pipe slightly below the springline, corresponding to the 4:00 and 8:00 positions on a clock face. For a typical PCCP section 20 feet (6.1 m) in length, 64 NDT traces were obtained. Eight of the traces, corresponding to when the four sensor array was positioned over the pipe joint, were not considered in the evaluation because of the discontinuity encountered as the signal passes from one pipe length to the next. Each NDT trace was examined and rated as either good or anomalous. A normalized NDT rating was calculated and entered into the spreadsheet for each pipe length by dividing the number of anomalous NDT traces by the total collected. Using this technique, a value of 0.0 indicates a pipe length with no NDT anomalies detected while a value of 1.0 would mean all of the readings were anomalous.

NDT Limitations

During the inspection and analysis of the 96-inch (2.4-m) PCCP water main, several physical pipeline features were found that interferes with the NDT signal and produce poor or invalid results. Pipe that was encased in concrete gave invalid results because the equipment does not distinguish the boundary between the outer mortar coating and the encasement pipe that is strapped in place with an annular void between the tunnel and the carrier pipe gave mixed NDT results. This however, is attributable to the proximity of the NDT sampling position to the top of the concrete cradle supporting the pipe. Good signals were produced when the test was taken above the concrete cradle, but they were disturbed when the array was opposite the cradle. Also, fittings or specials that were fabricated from welded steel and pipe lengths that were manufactured with a steel saddle for reinforcement of outlets, appeared to give invalid NDT data. Further research in the NDT signal trace interpretation may provide insight into the physical condition of more complex configurations of prestressed pipe.

Overall Pipeline Evaluation

Once all the information collected on the pipeline had been entered into the computer, an overall rating system was devised to allow rapid comparison of various pipeline rehabilitation options and their relative costs and risks. Through the application of a series of Boolean expression within the developed computer spreadsheet, a system was developed to first rate each piece of pipe from an anticipated structural performance perspective and then condition the analysis with land use considerations. Entering two factors into the program, one for pipe condition based on the NDT and the other accounting for land use, a list of which pipe sections would require rehabilitation or relining was produced based on the evaluation criteria selected. The results also calculated an overall planning level pipeline rehabilitation cost estimate that accounts for whether or not each pipe section was slated for major repairs. This approach allowed a rapid comparison of available rehabilitation options, ranging from minimum repairs to complete pipeline relining, along with their associated costs and risks.

Rating System to Account for NDT Results

An initial rating system was formulated to account for the condition of each pipe section that was based primarily on the NDT analysis results. Each pipe length was assigned a Base Evaluation Value (BEV), between 0 and 10 which corresponds to the normalized NDT rating multiplied by 10 and rounded to the next highest integer. Pipe sections within tunnels, encased in unreinforced concrete, and manufactured with steel outlet saddles was assigned a BEV of 3 to account for their unknown condition. A BEV of I was assumed for reinforced concrete encased pipe lengths due to the added strength the encasement supplies. Steel pipe specials were assigned a BEV of ° due to their relative low risk.

NDT Based Evaluation

The NDT results were used to select pipe in need of complete rehabilitation by comparing the BEV against a pre-selected maximum value. A BEV of 3, in the absence of visible or audible indications of distress, was selected as the maximum acceptable level of potential pipe distress. This limit was selected because the results of a previous investigation of approximately 4 miles (6.4 km) of 108-inch (2.74-m) PCCP sewage force main, conducted during the initial development of the NDT techniques, determined that an NDT factor of 0.4, which corresponds to a BEV of 4, was representative of a pipe with approximately five feet of broken prestress wire. It must however, be noted that the distressed 108-inch (2.74-m) pipe section was also found to have a large corresponding hollow in the inner concrete lining that was detected through manual sounding methodology. The significance of this is that a PCCP pipe length with a similar BEV would be indicative of considerable distress that has not progressed to the point of detectability without the use of NDT techniques. This is further supported by past experience with internal PCCP inspections at the Washington Suburban Sanitary Commission (WSSC) which indicate that, in the absence of certain metallurgical or environmental conditions, the pipe can be expected to provide reliable service for up to five years in the absence of significant internal cracking or concrete lining hollows. Scheduled inspections within this time period can be expected to detect severely distressed pipe prior to incipient failure.

Rating Land Use Considerations

Due to the potential of property damage and the risk of endangering the general public in the event of a pipeline rupture, land use criteria were developed to factor these concerns into the pipeline evaluation. A field survey of current land use features was conducted to update any changes since the pipeline was constructed. The approximate distances from any significant structures or otherwise potentially inhabited land features were measured from the pipeline construction plans and each piece of pipe was assigned to a Land Use Category (LUC) that corresponds to the relative level of risk. Five risk levels were defined based on the proximity of the pipeline to the land use features as defined in Table I that decrease from high risk to little or no risk. The LUC and the distance of significant land use features from the pipeline were entered into the computer spreadsheet for each pipe section.

Distance Rating Applied to LUC

The distance from each LUC was used to evaluate the relative land use risk in the event of a pipe burst. Three levels of risk were assigned and appropriate factors were applied in the analysis. The distance ranges selected, in order of decreasing risk, were 0 to 50 feet (0 to 15 m), 50 to 100 feet (15 to 30 m), and greater than 100 feet (30 m). If the pipeline should burst within 50 feet (15 m.), significant damage was predicted to a land use feature, while from 50 to 100 feet (15 to 30 m) away lesser damage was anticipated. At distances in excess of 100 feet (30 m), little or no damage to significant structures was assumed from a pipe rupture. Appropriate Land Use Adjustment (LUA) factors were then selected for each LUC as presented in Table I.1.

Table I.1

Land use category assignment (LUC)

Land Use Category (LUC)		Land Use Description	Land Use Adjustment (LUA) for Distance from Significant Features		
			0 to 50 feet (0 to 15 m)	50 to 100 feet (15 to 30 m)	Greater than 100 feet (30 m)
Decreasing Land Use Risk ↓	1	Inhabited Structures and Dams	5	4	0
	2	Major and Residential Roads, Railroads, and Public Transit Lines	4	3	0
	3	Parking Lots and Public Transit Repair Depots	3	2	0
	4	Other Roads (Rural Access, Remote Driveways)	2	1	0
	5	No High Risk Land Use	-2	-2	-2

Land Use Evaluation

The sum of the BEV and LUA were compared to a preset limit to select additional pipe lengths that present an unacceptable risk and will require rehabilitation or replacement. A maximum combined BEV and LUA of 6 was decided to be the highest acceptable for combined pipe condition and land use considerations. By combining pipe condition assessment with land use factors, an additional safety factor is applied in the analysis at locations where public safety could be at risk. At the other end of the spectrum, when a pipe has been rated as acceptable from a structural performance point of view is located away from a significant land use risk, a relatively larger risk of failure is acceptable.

Formulation of Rehabilitation Options

Three overall pipeline repair options were selected that differed only in which of the deficiencies and defects identified in previous investigations and reports were to be corrected. Repair Option 1 assumed that all identified items were to be corrected, Deficiencies that were identified as marginal but still serviceable were then deleted to produce Repair Option 2. Repair Option 3 was essentially the same as Repair Option 2, but was based on the assumption that the pipeline could be operated at a reduced internal pressure and thereby eliminate the need to upgrade thrust restraint for valves and fittings that were determined to be questionable. By varying the BEV from 1 to 3 and the LUA from 1 to 6, eighteen estimates were tabulated for each of the three repair options for a total of 54 rehabilitation choices. A graphical representation of the pipe lengths selected for rehabilitation or replacement, based on selected 'BEV and LUA values, was also produced that shows their relative location within the pipeline. The graphs also delineated, which sections were selected based on the NDT data alone and which were identified through, combined NDT and land use considerations. The options for rehabilitation of the 96-inch PCCP water main presented to WSSC management ranged from the minimum that was considered acceptable, from both performance and land use perspectives, on up to complete relining with welded steel.

Results and Conclusions

The application of the PCCP evaluation methods presented have altered WSSC's previous plan to reline the entire 96-inch (2.4 m) PCCP water main with welded steel, at an estimated cost of over 52 million dollars, to produce a reliable pipeline. A managed PCCP rehabilitation approach, with a projected cost of 20 million dollars, has been selected instead. This corresponds to Repair Option 2, a BEV of 2, and a combined BEV and LUA of 6 or less. This approach allows the WSSC to extend the service life of their existing capital investment and maintain reliable service to consumers while still addressing public safety issues related to high-density development. Future refinements of the inspection and NDT techniques employed, in conjunction with improvements to the evaluation logic used on the 96-inch (2.4-m) PCCP water main, can be expected to yield even greater savings on future projects

CASE STUDY 3: VALIDATION OF PCCP PIPELINE NON-DESTRUCTIVE TEST RESULTS

Manufacturer/Patent holder : PPIC
Assessment technology : RFEC/TC

Ortilano Oligo, Water Supply & Treatment Superintendent, City of Aurora, CO; 18301 East Quincy Ave., Aurora, CO 80015; phone 303-690-4977; boligo@ci.aurora.co.us
Larry Catalano, P. E., Farnworth Group, 2696 South Colorado Blvd., Suite 250; Denver, CO 80222; phone 303-692-8838 lcatalano@f-w.com
Will Worthington, P. E.; Pressure Pipe Inspection Company, 1435 N Hayden Rd, Scottsdale, AZ 85257; phone 480 990 2466; (at the time of this project: President, Pipeline Technologies, Inc.; Scottsdale, AZ) will@pipetech.com

Introduction

The City of Aurora, Colorado, is a city of more that 250,000 residents, which lies in the Denver Metropolitan Area. The City has established a mission for its Water Operations, which is simple and straightforward: "Provide safe and dependable water services for the citizens of Aurora".

Based on their established values of Customer Service, Teamwork, Integrity, Excellence and Innovation, the Water Operations team strives to more than just good; their collective vision is to become a recognized leader in the Water Industry.

The City supplies water to residents in its service area from an extensive water supply system whose major source is the South Platte River. Water is diverted from the South Platte River at the Strontia Springs Reservoir and conveyed through a series of tunnels and reservoirs to Rampart Reservoir. The Rampart Parallel Raw Water Transmission Pipelines deliver water to the City of Aurora water treatment facilities from Rampart Reservoir, a distance of 22 miles (35.43 km). A portion of the line is constructed of 54-inch (1.37-m) pre-stressed concrete cylinder pipe (PCCP) manufactured by United States Pipe and Foundry in 1980. This portion of the line has been the subject of investigation by the City for several years due to their discovery of distressed pipe in this reach, and to their concern regarding reliability of PCCP. The City of Aurora Collection System is illustrated in Figure I.1.

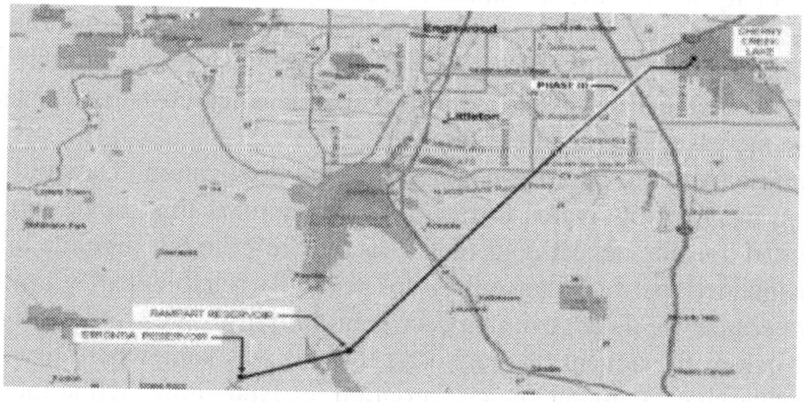

Figure I.1 City of Aurora Collection System (*Source*: Oligo, O., Catalano, L., and Worthington, W., *Validation of PCCP Pipeline Non-Destructive Test Results*, 2003)

As part of their investigation, the City removed the line from service and performed an internal inspection of this pipeline in February and March 2000, using Remote Field Eddy Current/Transformer Coupling (RFEC/TC) techniques. This inspection was performed by the Pressure Pipe Inspection Company, and indicated the presence of broken prestressing wires in 111 sections of pipe out of the 1007 pipes inspected. Three of these pipes were judged to have 20 or more breaks in the prestressing wire. The section that exhibited the most severe distress was #959, which showed evidence of 40 breaks in prestressing wires. The adjacent pipe #958 showed evidence of 15 breaks of wire.

There have been no catastrophic ruptures of this pipeline, although the City did discover one distressed area several years ago. The RFEC/TC test results were cause enough for concern, however, because of the importance of the pipeline to the City's water deliver system, and the setting of the distressed pipe in a commercial office complex area where collateral damage of rupture would be extensive. The City removed and replaced those two sections of pipe then performed an "autopsy" on the worst pipe section to validate the results of the RFEC/TC test, and to identify any obvious causes of any damage found to be present.

The Rampart Pipeline Phase III is constructed of Prestressed Concrete Cylinder Pipe (PCCP), which was developed and used in the United States in the mid-1940's as a means to provide an economical means of conveying water under pressure. The nature of the steel alloy used and the method of applying the steel wire to the concrete pipe core under tension, made very efficient use of the concrete and steel materials that constitute PCCP. The competitive cost of PCCP, its durability, and its anticipated long life led to increasing acceptance and use of the product. By the late 1960's, PCCP was the preferred pipe product for large pressure water pipelines in the United States. The product is widely used in other parts of the world as well, including Europe, the Middle East, Australia, South America, North Africa, and China.

Due to a wide variety of causes, there are over five hundred recorded failures of PCCP in the world. The causes of failure may be related to improper design, manufacturing flaws, improper construction, improper operation, environmental conditions, and damage by outside agents. In many instances, it is difficult to assign a specific cause to a particular pipeline, and indeed this alludes to another problem faced by pipeline owners. Regardless of the causes of pipe distress, the fact is that our assets are continuing to age, and failures will occur with increasing frequency as they become older. In the absence of tools to diagnose the condition of any asset, it is virtually impossible to maintain the asset; and its catastrophic failure may occur without warning. On the other hand, given a reliable means to diagnose pending failure, and the tools to rehabilitate at these points, the asset will be maintainable. It will take on a new life, as the owner is able to manage the asset, to project maintenance costs, and achieve a maximum useful life with an increased degree of confidence.

The American Water Works Association's recent publication "Dawn of the Replacement Era – Reinvesting in Drinking Water Infrastructure," captures this theme with a great deal of clarity. This "buried treasure beneath our feet" – as they refer to our pipeline infrastructure – has not been well maintained, partly because they are generally reliable, partly because we can't see them, and partly because we didn't even pay for them initially. We would add another reason to our neglect, that before the advent of diagnostic tools it was somewhere between difficult and impossible to undertake a maintenance program. And now with the introduction of electromagnetic and acoustic tools in the toolbox, many new possibilities become apparent. PCCP maintenance is analogous to health care or automobile maintenance or other pipe materials for that matter, in several respects. Physicians and mechanics alike depend on diagnostic tools. Pipeline owners are developing the same capability. We need to develop the confidence in them before we are confident, however. Do they work?

Results of Investigations

The City of Aurora had performed the electromagnetic pipe tests in early 2000. Pipe #959, the "worst" section of pipe, was excavated and visually inspected on January 31, 2001. Initially there was no indication of distress of either Pipe #958 or #959. On closer examination, a 3-inch (75-mm) diameter corrosion stain was observed at the 8:30 position about one foot from the north (or downstream) end of the pipe section. This area was also revealed to be slightly "drummy" upon tapping with a hammer. The corrosion of the wire, the expansion of corrosion products, and subsequent de-lamination of the mortar and core of the Pipe #959 caused this. The mortar was then removed from an area of about one square foot, revealing the presence of several breaks in the prestressing wire. There was no other evidence of distress in Pipe #959 at this point in the investigation. See Figure I.2.

Following the excavation of the two sections of pipe on January 31, 2001, pipe #958 was cut into sections to permit its removal from the bell-and-spigot pipe sections. This allowed the removal of Pipe #959 without damage. Both sections were transported to the City storage yard for further investigation. These sections were replaced in the pipeline with steel sections, and the pipeline was returned to service several days later.

There is no existing technology to nondestructively identify the presence of corroded wire beneath the mortar coating. This has been a continuous problem in pipe testing and failure investigations. Until such a method is available, the only method to determine the condition of the wire is to destroy the pipe and visually inspect the wire.

An "autopsy" was performed on Pipe #959 by Pipeline Technologies, which entailed the dissection of the Pipe #959 by removing of the protective mortar and inspection of the condition of the prestressing wires. Samples were taken of two of the most important components of the Pipe #959, namely the prestressing wire and the mortar that protects the wire. Measurements of all pertinent dimensions of the Pipe #959 were taken.

Figure I.2 Initial investigation (*Source*: Oligo, O., Catalano, L., and Worthington, W., *Validation of PCCP Pipeline Non-Destructive Test Results*, 2003)

The removal of the mortar was begun by making an incision (Figure I.3) through the mortar and prestressing wire at the 12:00 position, from end to end of Pipe #959. After this incision, the mortar and wire were manually separated from the entire section of pipe, observing and recording the condition of the mortar and wire. The Pipe #959 was at this point resting on two timbers a few inches above the ground, and the mortar could be removed to about the 5:00 and 7:00 positions. We had the opportunity to roll the Pipe #959 over 90 degrees so that its 3:00 position was resting on the ground. This then allowed the complete removal and inspection of the mortar and wire from the entire circumference of the Pipe #959.

Figure I.3 Autopsy begins with incision (*Source*: Oligo, O., Catalano, L., and Worthington, W., *Validation of PCCP Pipeline Non-Destructive Test Results*, 2003)

Figure I.4 Top of pipe shows extensive distress (*Source*: Oligo, O., Catalano, L., and Worthington, W., *Validation of PCCP Pipeline Non-Destructive Test Results*, 2003)

It is noteworthy that there was no corrosion present on the bottom of the Pipe #959. This is consistent with other investigations, however this investigation differed from many others in that there was the opportunity to strip the mortar and carefully examine the bottom portion of the pipe. In most instances, investigators have only glimpses of the bottom and note an apparent lower rate of corrosion. There is no scientific explanation for the lower rate of corrosion on the bottom half of the pipe. The most plausible explanation for this is that the top half of the pipe is subject to cycles of wetting and drying as a result of downward percolation of precipitation through the soil. It is well known that wetting and drying of concrete does tend to accelerate corrosion of reinforcing steel, whereas a constant moisture level is not corrosive. The theory holds that the bottom of the pipe is shadowed from the percolation and therefore does not experience the moisture cycles and therefore does not corrode as readily. See Figures I.4 and I.5.

Figure I.5 Bottom of pipe is free of distress (*Source*: Oligo, O., Catalano, L., and Worthington, W., *Validation of PCCP Pipeline Non-Destructive Test Results*, 2003)

Figure I.6 Distress area with 8 broken wires (*Source*: Oligo, O., Catalano, L., and Worthington, W., *Validation of PCCP Pipeline Non-Destructive Test Results*, 2003)

In summary, fifteen separate areas of corrosion were identified. Eleven of these areas contained one or more broken wires, and the total number of broken wires was fifty-two.

It is also noteworthy that the measures now routinely taken to provide electrical continuity throughout the pipeline were not taken in the manufacture of Pipe #959. Bonding straps were not placed beneath the prestressing wire at the time of manufacture of Pipe #959. Neither were there any shunting straps between the adjacent pipe sections. Presumably the entire pipeline was manufactured and constructed the same way, although there are instances when these straps are inadvertently omitted during the manufacture of a section of pipe. This is not inconsistent with manufacturing practice in 1980 when Pipe #959 was made. The absence of these measures to provide electrical continuity does mean that it will be virtually impossible to provide cathodic protection to the pipeline to reduce the rate of corrosion.

Tests were performed on the mortar, prestressing wire, and soil surrounding the pipe to gain some insight into the cause of distress at this location. The wire tests included the torsion test, the ammonium thiocyanate (NH_4SCN) test, and mechanical tests. These tests indicate that the prestressing wire used in the manufacture of Pipe #959 was of inferior quality, as evidenced by the signs of severe strain aging of this wire during the manufacturing process (Figures I.6 and

Figure I.7 Distress area with 12 broken wires (*Source*: Oligo, O., Catalano, L., and Worthington, W., *Validation of PCCP Pipeline Non-Destructive Test Results*, 2003)

I.7). The most serious consequence of strain aging is the formation of longitudinal cracks and splits. These flaws in the longitudinal direction do not reduce the load carrying capacity of the wire initially, but they do make the wire difficult to protect with the mortar coating, and the wire is therefore prone to corrosion and failure in that the porosity of the mortar is below standards. The mortar was also judged to be inferior in that it was not dense and durable as required, and it was applied in a manner that did not encapsulate the entire wire and caused some separation in the mix. The inferior protective mortar, combined with split prestressing wire that is difficult to protect under the best of conditions, will inevitably lead to early failure of a pipe section. These conditions apparently caused the distress in these pipe sections.

Conclusion

The pipe autopsy confirmed that Remote Field Eddy Current/Transformer Coupling nondestructive testing of the pipeline is effective. Prior to the RFEC/TC test, the City had no reason to suspect the presence of highly distressed pipe at this location. The RFEC/TC test indicated the presence of a structurally significant number of broken wires (40 wraps), and while the actual number of breaks was 52, this discrepancy does not alter the overall effectiveness of the method. Confirmation of the effectiveness of this test method has served two important purposes. First, it confirmed the wisdom of removing and replacing this section of pipe before it failed in an expensive and damaging rupture. Secondly, it enabled the City to gain enough confidence in electro-magnetic testing that it can be incorporated into a program of diagnostic condition assessment that promises to add decades of useful life from this facility, while improving the reliability of water deliveries and avoiding the cost of catastrophic ruptures.

CASE STUDY 4: LIFE PREDICTION AND CONDITION ASSESSMENT FOR LARGE DIAMETER WATER PIPELINES

Manufacturer/Patent holder	:	Advanced Engineering Solutions Limited
Assessment technology	:	MFL

Introduction

Advanced Engineering Solutions Ltd. (AESL) inspect and assess the condition of low-pressure pipelines and associated structures for a wide range of utility industry clientele, particularly water companies. Pipeline condition assessment is a major part of the ASEL business and other activities include high level pipe system studies, pipe system maintenance, structural renovation and corrosion protection. This specialization in pipeline asset management services has required the development of inspection technologies, data analysis techniques and site to office communications to refine the condition assessment process and has introduced the basis of risk management for strategic low pressure pipelines. This paper considers AESL's approach to pipelines condition assessment and operational life predictions for large diameter pipelines.

Utility Pipelines

Low-pressure pipelines can be defined as those, which may be critically stressed by internal (pressure) or external (ground) loading regimes or any combination thereof. Condition assessment must include these potential loading combinations and their influence on pipeline integrity, as well as recognizing any pipe wall deterioration. When developing maintenance and replacement strategies for older pipelines, the importance of past performance as an indicator of future capability must also be recognized. The large numbers of small diameter cast iron pipe failures in the utility industries allows performance history to be used as a basis for replacement strategy, although the U.K. water industry also uses local inspections of these pipelines as input. AESL carry out the inspection process on the full range of pipe diameters, although the full condition assessment activities focus on pipelines of 12-inch (400 mm) diameter or greater.

For large diameter pipelines, there are generally insufficient failures on which to base an entire replacement strategy and the approach necessarily taken is to develop an understanding of their current structural performance capability. Typical utility pipelines will vary substantially in age, but experience shows that the operational life achieved is usually most dependent on the pipe material, pipe manufacturing processes, and the environment local to the pipe. Some pit cast gray iron pipes continue to provide operational service without significant problems after more than 100 years, whilst more modern materials, typically spun gray and ductile cast iron and thin wall steel, can deteriorate more rapidly to an unacceptable condition.

Older pipeline materials of interest to the water industry include steel, pit and spun cast gray iron, ductile iron, PVC, GRP and concrete. AESL typically carry out condition assessment on metallic pipes, at all sizes and pressures. Condition assessment processes can be applied to pipelines, pipe systems and associated structures in almost any location. However, it is common for pipeline operators to focus their attention towards strategic sections of a pipeline, which may be critical, in either an operational or technical sense, to providing or maintaining supply to customers. Specific locations may also be strategic or critical, where pipelines cross over, or under, roads and railways, rivers and canals, areas of rough terrain, marshland, etc and the ultimate pipeline capability is dominated by the performance of supporting structures. Pipeline maintenance strategies based on asset condition monitoring now increasingly direct the inspection and assessment process to specific sections and locations that are strategic or critical; see Figure I.8.

Pipeline Asset Management

Regulation of U.K. utility companies has resulted in pipeline managers striving for a better understanding of the financial values and optimum performance capabilities of their pipelines and related assets, to allow maintenance and capital expenditure to be prioritized. Variations in design, construction, operating and maintenance standards of pipelines, particularly older pipeline's, make meeting this requirement challenging. AESL's interpretation of the asset management process is shown in Figure I.9.

Figure I.8 Typical "critical" pipeline bridge crossing (*Source*: Wayman, M., Risk Assessment in the Pipeline Industry, 2003)

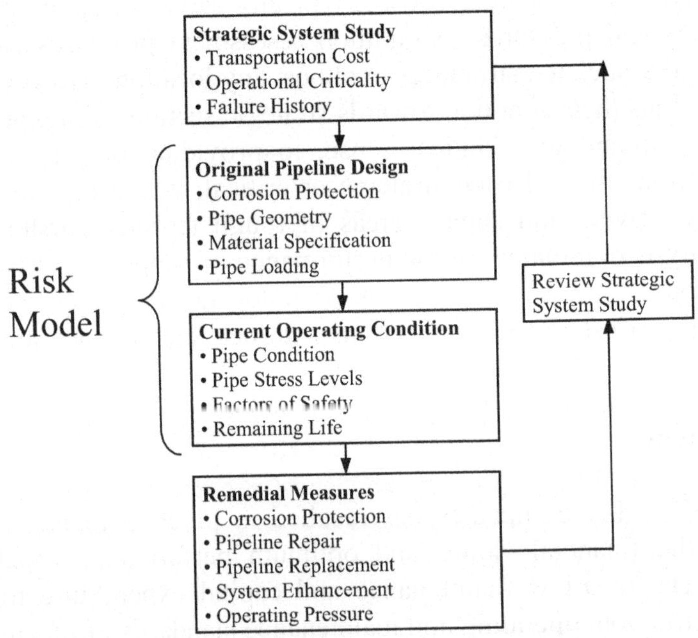

Figure I.9 Pipeline asset management process (*Source*: Wayman, M., Risk Assessment in the Pipeline Industry, 2003)

AESL's asset management process has four distinct stages:

- A strategic system study, which requires significant pipeline manager input, focussed on developing a sound understanding of the pipeline, its operational requirements, in-built redundancies and operating costs.

- A two stage risk modeling process which requires an understanding of:
  - The original pipeline design parameters including, if possible, envisaged loading regimes and design performance capability.
  - The pipeline's current performance capability, based on a recent pipe wall inspection and knowledge of the historical and current loading regimes. This process may also allow the remaining operational life to be predicted.
- Appropriate remedial measures are identified after the inspection, condition assessment and risk modeling processes. This commonly involves refurbishment of strategic sections to improve the performance capability of the entire pipeline.
- Finally, after risk modeling and remedial work, a further review of the strategic system study allows changes in criticality between different pipelines or pipe sections to be recognized and new targets set for future condition assessment, risk modeling and remedial work.

The complete asset management process provides a strategic review of an entire pipe network, to identify specific pipelines or pipe sections, which are critical in terms of the potential for, and consequences of, a major failure. This review could be a full analysis of the system design, materials and operating regimes. However, it is generally the case that selection of 'critical'ocations can be based on the pipe system manager's understanding of previous failure issues, supply criticality etc. In terms of optimizing the pipe system management and maintenance process it is necessary to develop an understanding of the pipeline's operating life.

Pipeline Life Modeling

A 'bath tub' curve is a good basis for understanding a pipeline's life cycle (Figure I.10). The curve provides a reasonable representation of the performance capability of many engineering structures, including pipelines, and outlines the life cycle of the pipe or pipe sections under predicted loading regimes and deterioration mechanisms. However, only 'natural' pipeline failures are represented, not failures due to unpredictable extremes of loading or third party interference. The curve illustrates a pipeline service life in three phases with differing failure rates:

- **Premature Failures** can occur at a higher than expected rate soon after a pipeline is constructed, due mostly to design and construction problems or extreme environmental circumstances not recognized when the pipeline was designed.
- **Pipeline Service Life** is the true working life of a pipeline, which should be characterized by no failures due to design errors or to deterioration of the pipeline under the predicted loading regime.

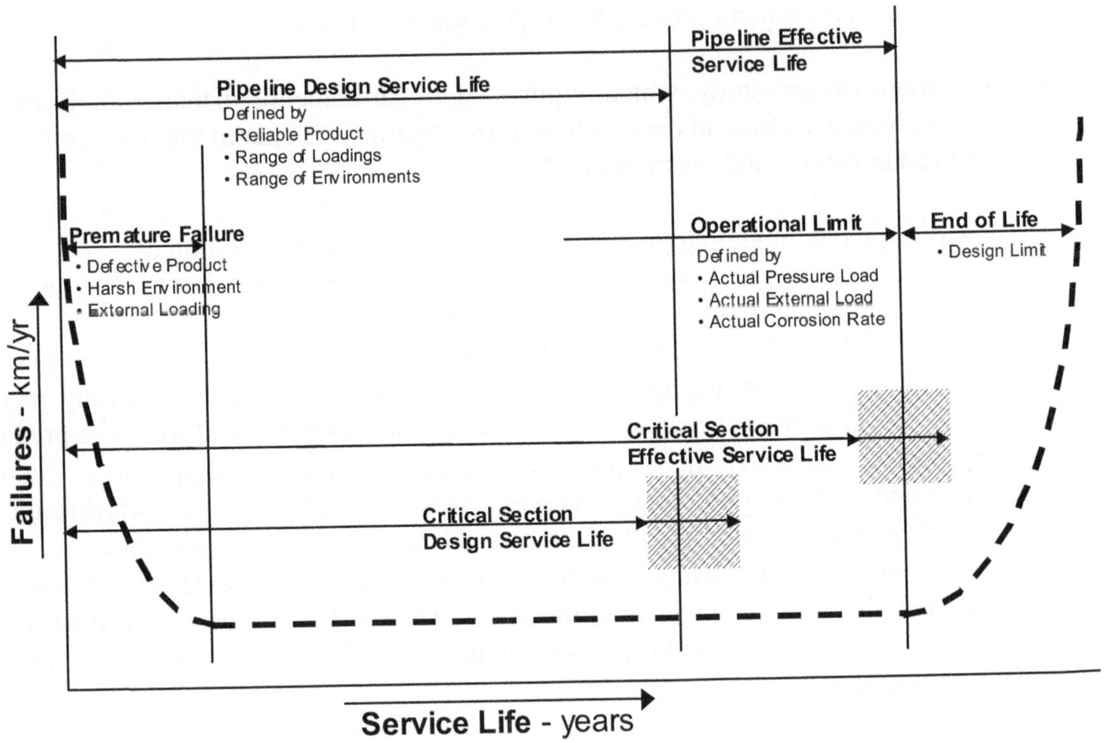

Figure I.10 Life cycle of typical pipeline (*Source*: Wayman, M., Risk Assessment in the Pipeline Industry, 2003)

- **End of Life** is the point where the number of pipeline failures, due to structural inability to withstand applied loads, starts to become problematic. Defining the 'End of Life' point requires consideration of technical and economic issues, including safety, and crucially depends upon the operator's understanding of the pipeline system and its required performance.

It is a premise of the approach developed by AESL that large diameter low pressure pipelines should be considered as operating within the 'pipeline service life' section of the curve and that pipelines operating beyond their natural service life should have been or can be recognized as such. Generally, premature failure issues on in-service pipelines will have been resolved many years ago and significant numbers of failures due to pipe wall deterioration will not yet have begun to occur, or if occurring will be recognized as resulting from this cause.

Within this model, the deterioration process can be considered to be related to the pipe material, corrosion protection and environment. These issues are complex. However, by carrying out local inspection and optimizing both the inspection process and the outputs using statistical techniques, comparable condition assessment outputs can be targeted. A typical natural life prediction for spun cast iron pipe, based on sensible predicted corrosion rates, is shown in Figure I.10, illustrating a typical pipe material and the significant influence of diameter and wall thickness on the theoretical natural life.

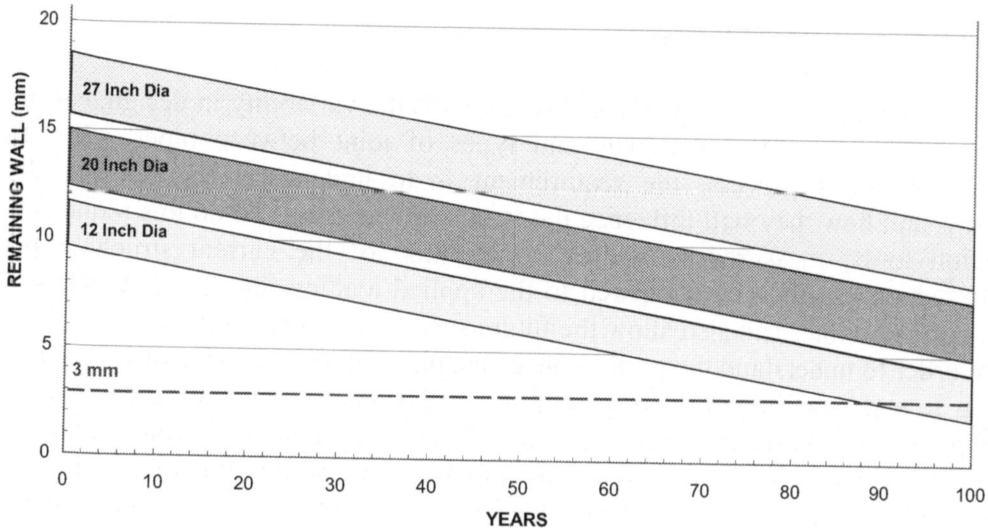

Figure I.11 Natural corrosion of spun cast gray iron pipe (BS 1211) (*Source*: Wayman, M., Risk Assessment in the Pipeline Industry, 2003)

Operating Life of Pipeline

The operational service life of a pipeline can be considered as comprising two main components, the *Design Service Life* and *Effective Service Life*; see Figure I.11. The difference between these two components can represent a usable extension in the overall pipeline life and condition assessment and risk modeling processes are targeted at assisting the management of this extended life.

The *Design Service Life* is the life that was expected when the pipeline and sealing arrangement, pipe and other components, etc. were originally designed, all based upon expected loading regimes, environmental protection and pipe wall deterioration. In practice, many pipelines can operate well past the end of the *Design Service Life*, due to fundamental redundancies built in to the pipeline at the design stage.

The *Effective Service Life* is the period between the defined Design Service Life and the point at which the pipeline can be considered as no longer meeting the necessary structural performance capability. There are sound practical reasons why the Effective Service Life may be longer than the Design Service Life, i.e. the pipe wall thickness may be greater than the minimum allowed, effectively increasing the pipe strength and its corrosion allowance. Also, pipelines are often operated at less than the maximum design pressure, significantly reducing the applied stresses.

Technically, the *End of Life* can be defined as the point in time when the pipe wall has thinned to a level where accurate predictions of the structural integrity under the applied loading are no longer possible. Economically, the *End of Life* can be defined as the point at which the cost of repairing localized pipe failures, which can be predicted based on current structural strength and applied loading, is such as to make replacement or large scale refurbishment economically attractive. Neither the *Design Service Life* nor the *Effective Service Life* are necessarily fixed. There is usually a significant variability in loading regimes and corrosion rates, which can allow the pipeline manager to maximize the life of the pipe system before making significant capital investment.

The Condition Assessment Process

An important issue with pipeline assessment is the variability in design, construction and maintenance standards, pipe materials and types of joint between pipes. In terms of the condition assessment process, the requirement is to understand the relevant deterioration mechanisms and how they will influence the structural integrity. To do this requires the level of deterioration to be measured, providing a definition of the current structural performance capability, which can then be compared to the applied loading regime. Estimating the overall effective corrosion rates can also allow the future service life to be predicted.

In order to understand the performance capability of an in-service pipeline, it is necessary to develop an understanding of the original pipeline design, original and current loading regimes, current pipe wall condition and the effectiveness of any environmental protection measures. It is also necessary to identify locations where the pipeline is most highly loaded and, ultimately, the Factor of Safety is at its lowest. From the customer's point of view, these highly stressed locations may not coincide with the most operationally critical locations. Operational criticality is general related to safety, economic and environmental issues. AESL have developed condition assessment processes for pipelines, applying existing technology where possible, and developing further technologies where necessary.

The ongoing deterioration of a pipeline is the major factor determining its current performance capability and remaining life, requiring an accurate assessment of the current pipe wall thickness by physical inspection. Ideally, this would involve a full pipe wall inspection of the pipeline or pipe section, but this is not generally practical for both economic and technical reasons. Selective inspection is therefore used to provide appropriate data on pipe wall condition, supported by the use of stress analysis and statistical techniques to focus the inspection towards the most appropriate positions or section along the pipeline length.

Pipeline Inspection

Selective inspection can provide the required numerical data, which can be validated using appropriate statistical techniques. The magnetic inspection tools developed and applied by AESL are designed for external use on the pipe where access is possible or where access can be provided by excavation. Variations in the pipe magnetic properties may require tools to be calibrated for specific pipe materials using conventional ultrasonic techniques. The purpose designed inspection tools allow 'mapping' of internal and external corrosion in the pipe wall. The tools are designed for easy fitting to pipelines, tools of 12-inch (400 mm) diameter and greater being used to inspect sequentially around the diameter of the pipe of any diameter. Full circumferential inspection of the small diameter pipe is carried out.

The resulting magnetic inspection data can be imaged directly, providing the customer with both defect size data and an image to which the dimensional data can be related; see Figure I.12. Brief technical literature for both small and large diameter inspection tools is provided below.

For large diameter pipes, it is necessary for the structurally important defects to be defined as part of the condition assessment process. For high-pressure pipelines, effectively working as pressure vessels, defect sizes considered to be important are well defined, based upon fracture mechanics. However, for large diameter low pressure pipelines, a range of defects can be considered critical, depending on internal pressure, external loading regime and the way in

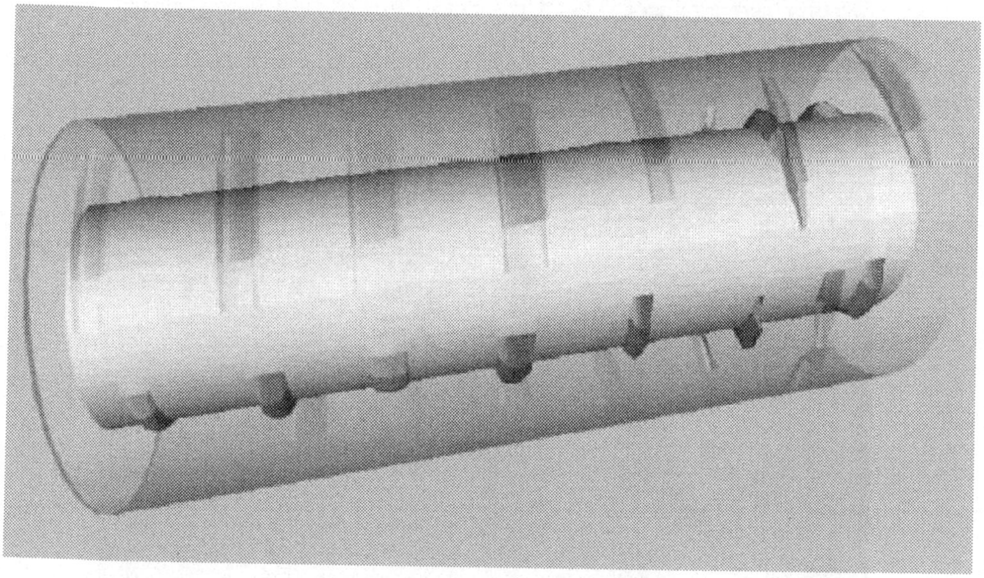

Figure I.12 Imaging of inspection outputs (*Source*: Wayman, M., Risk Assessment in the Pipeline Industry, 2003)

which the pipeline is supported. The scale of deterioration, which potentially would lead to failure, will also vary with pipe diameter, wall thickness, loading regime etc. The condition assessment process must therefore consider the criticality of a range of defects and focus inspections towards the most critical. A major influence on pipe deterioration is the type and quality of corrosion protection originally applied to the pipe and how it has been maintained during its service life. AESL base their approach to understanding the risk of failure on a significant knowledge of pipeline technology, failure mechanisms and inspection outputs.

The variability in design and condition of pipelines encountered in the field requires a strong coordination between field inspection teams and the engineers carrying out the stress analysis and statistical validation. To facilitate this coordination, and optimize the services provided to pipeline operators, AESL now utilize "Blue Tooth" technologies which allow inspection data to be transferred directly from tools to hand held computers and thence to AESL's main computer systems; Figure I.13. Inspection output data can be analyzed on site, allowing field inspection teams to have an immediate understanding of the pipe wall condition and provide rapid feedback to the customer. This wire-less approach also improves operational safety when working on pipe bridge crossings, tunnels etc.

Analyzing pipeline performance, particularly if focussed towards critical sections, allows deterioration issues to be identified and resolved. A broader requirement is to be able to compare the performance capability and medium to long-term maintenance requirement of individual pipelines within a wider pipe system. For these purposes, a standardized risk modeling approach is required. However, risk modeling for low pressure pipelines can be complex, due to wide variations in original design and construction standards, pipe materials, loading regimes, etc, and changes to these parameters over time. The main requirement is to determine the probability of pipe failure and the likely consequences, and compare them in a mathematically robust way. AESL is currently developing an approach to risk assessment

Figure I.13 Small diameter inspection tool downloading data (*Source*: Wayman, M., Risk Assessment in the Pipeline Industry, 2003)

modeling, based on an understanding of the performance capability of a particular pipeline and incoming condition data for critical sections of that pipeline from field inspection teams.

AESL Condition Assessment Activities

AESL have completed, and have ongoing condition assessment projects for many of the U.K. water companies, and also carry out condition assessment activities for the U.K. gas industry. These projects range from the high level analysis of pipe system failures to allow inspection, condition assessment and ultimately maintenance requirements to be defined, to the detailed inspection and structural performance capability analysis of critical pipe sections. The Water Industry condition assessment projects can be divided into those for small diameter pipe and related to water quality issues, and those for larger diameter pipes and related to system performance, valuation and maintenance planning issues.

In the January to December 2002 period the Company carried out a total of 100 large diameter condition assessment projects, and some 600 small diameter pipe inspections for U.K. water companies.

The following U.K. Water Companies are customers of AESL:

- Essex and Suffolk Water
- Northumbrian Water
- Scottish Water
- Severn Trent Water
- Southern Water

- Sutton and East Surrey Water
- Thames Water
- Welsh Water

AESL has also carried out work for the following non-U.K. water company:

- Dublin Corporation

AESL has also carried out work for the following consultancies on water pipeline engineering:

- AwwaRF (American Water Works Association Research Foundation)
- FaberMaunsell
- Montgomery Watson Harza
- NWTC

Conclusions

Condition assessment and life prediction for low-pressure pipelines is a complex process due to the wide variability in their design, construction, loading regimes and environment. However, there are many such pipelines in service and they are of substantial asset value. AESL are applying, and refining, condition assessment processes based on life prediction models, to provide methodologies for understanding these assets and managing their service life to the benefit of the pipeline manager and his customers.

ECAT – External Condition Assessment Tool

Overview

AES are able to announce a further addition to their wide range of inspection tools to support the pipeline assessment services offered by the company.

The External Condition Assessment Tool (ECAT) has been developed for rapid in-situ assessment of the pipe wall condition of iron and steel pipelines. The ECAT evaluates the level of corrosion in a pipe by measuring pipe wall thickness. The inspection unit determines corrosion levels, both internal and external at any location on the pipe using non-intrusive techniques (Figure I.14).

Units have been designed for operation in the full range of pipe sizes, from 3-inch to 48-inch (75 mm to 1200mm) diameter. Where access is permitted, 360-degree inspection can be carried out along any pipe length.

Figure I.14 ECAT – External Condition Assessment Tool (*Source*: Dickson, D.T., ECAT, 2000)

System Description

The ECAT inspection system is based on conventional magnetic flux leakage (MFL) inspection technology. MFL is the mainstream technology used by the oil and gas industry for internal inspection and assessment of iron and steel pipes.

The high flux magnetic inspection system does not normally require the protective coating on the pipe to be removed, and to carry out the inspection, minimum external preparation is required. The unit is simply placed on the pipe and after powering up, traversed along the pipe to capture the data. By use of multiple sensor arrays, full 360-degree coverage of the pipe can be obtained with the minimum number of sweeps.

Data from the system is captured and recorded on a laptop computer and displayed in 3D graphical and numeric form. Rapid on-site analysis of the data can be carried out to provide an immediate assessment of pipe condition (Figure I.15).

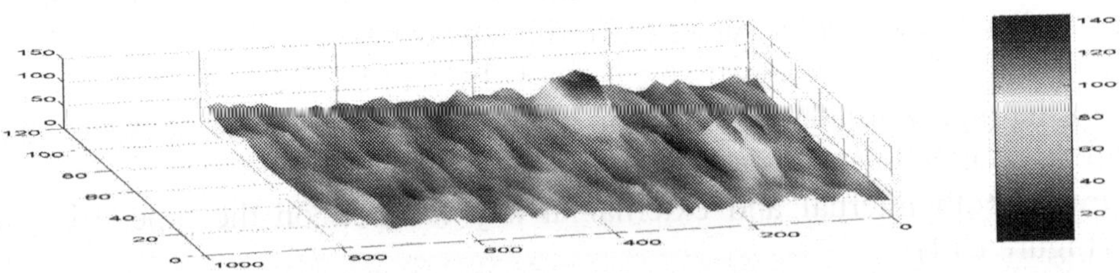

Figure I.15 3D graphical representation of corrosion pits in iron pipe (*Source*: Dickson, D.T., ECAT, 2000)

Figure I.16 S19 – External Condition Assessment Tool (*Source*: Dickson, D.T., ECAT, 2002)

S19 – External Condition Assessment Tool

Overview

The S19 Condition Assessment Tool (Figure I.16) allows rapid non-intrusive assessment of the condition of iron or steel pipes, to aid refurbishment or replacement decisions. Internal and external corrosion is identified and measured.

Tool Design

Tool operation is based on measuring magnetic flux leakage, and tools are available to allow inspection of pipe of 3-inch to 12-inch (75-mm to 300-mm) diameter.

Minimal external cleaning of the pipe barrel is required prior to inspection. The number of inspection sweeps required depends upon the pipe diameter:

- Full circumference wrap-around tools for 3-inch, 4-inch and 6-inch diameter pipe.
- Segmented inspection is used for pipe of 8-inch diameter and greater.

An on-board electronics unit records and stores the inspection data. At the end of a run this data can be downloaded to a PC and displayed in numeric and graphical form.

A 3D image of the remaining material can be generated as an aid to visualization of the condition of the pipe as shown in Figure I.17.

Inspection Performance

The inspection capability of the tool is focused on matching that achieved by grit blasting and measurement of pipe samples removed from the pipe.

The tool is quick to use, and inspection data can be checked and validated at the trench side.

The tool measures the remaining metal in the pipe wall and differentiates between internal and external corrosion. Corrosion patterns and individual defects can be identified and sized.

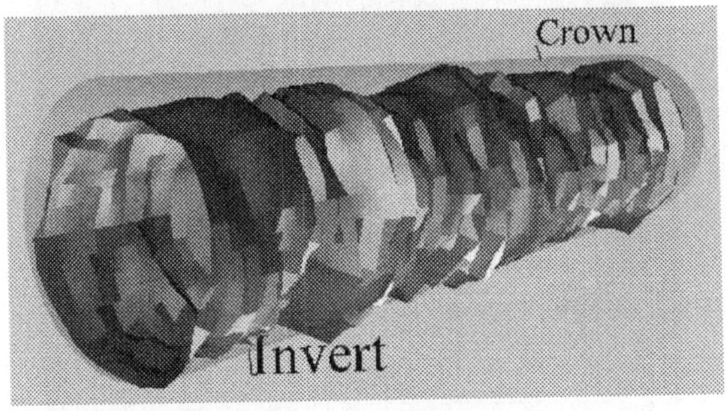

Figure I.17 3D graphical representation of pipeline structure (*Source*: Dickson, D.T., ECAT, 2002)

Inspection Outputs

The data collected can be used to automatically populate an inspection report, formatted to the customer's specifications, which would include sizing of the most significant external and internal defects as well as prediction of remaining life.

APPENDIX J
MANUFACTURERS AND SUPPLIERS

Manufacturer/Patent holder:	Rock Solid Pty. Ltd.
Address:	11 Evans Street Burwood, Victoria Australia 3125
Tel:	+61 3 9888-0033
Fax:	+61 3 9888-0077
Email:	info@rocksolidgroup.com.au
Web:	http://www.rocksolidgroup.com.au/
Technology:	BEM

Manufacturer/Patent holder:	The Pressure Pipe Inspection Company
Address:	4700 Dixie Road Mississauga, Ontario, L4W 2R1 Canada
Tel:	+1 (905) 624-1040
Fax:	+1 (905) 624-4777
Email:	info@ppic.on.ca
Web:	http://www.ppic.on.ca/
Technology:	RFEC/TC

Manufacturer/Patent holder:	Hydroscope Inc., USA
Address:	1776 Montano Blvd. NW Bldg #2 Albuquerque, NM 87107 U.S.A.
Tel:	+1 (505) 345-1897
Fax:	+1 (505) 345-1896
Email:	Info@hydroscope-USA.com
Web:	http://www.hydroscope.com/
Technology:	RFEC

Manufacturer/Patent holder:	NDT Corporation
Address:	67 Millbrook Street Worcester, MA 01303 U.S.A.
Tel:	+1 (508) 754-0417
Fax:	+1 (508) 754-0418
Email:	paul.fisk@ndtcorporation.com
Web:	http://www.ndtcorporation.com/
Technology:	Seismic Pulse Echo

Manufacturer/Patent holder:	Advanced Engineering Solutions Ltd.
Address:	South Nelson Road South Nelson Industrial Estate Cramlington, NE23 1WF United Kingdom
Tel:	+44 (0) 1670 739999
Fax:	+44 (0) 1670 717999
Email:	enquiries@aesengs.co.uk
Web:	http://www.aesengs.co.uk/
Technology:	MFL

Manufacturer/Patent holder:	Pure Technologies
Address:	3rd Floor, 705 11th Avenue SW Calgary, AB, T2R 0E3 Canada
Tel:	+1 (403) 266-6794
Toll free	1 800 537-2806
Fax:	+1 (403) 266-6570
Email:	mailbox@soundprint.com
Web:	http://www.soundprint.com/
Technology:	Acoustic Emission, Electromagnetic

Manufacturer/Patent holder:	WRc Sahara
Address:	Frankland Road Blagrove Swindon, SN5 8YF United Kingdom
Tel:	+44 (0) 870 607 1200
Fax:	+44 (0) 179 386 5001
Email:	sahara@wrcplc.co.uk
Web:	http://www.wrcplc.co.uk
Technology:	Acoustic Leak Detection

Manufacturer/Patent holder:	Guided Ultrasonics Ltd.
Address	17 Doverbeck Close Nottingham, NG15 9ER United Kingdom
Fax:	+44 (0) 1623 491093
Email:	info@guided-ultrasonics.com
Web:	http://www.guided-ultrasonics.com
Technology:	Guided Wave

REFERENCES

AMG (Applied Magnetics Group). 2002. Remote Field Through Wall Inspection Technique [Online]. Available:<http://www.physics.queensu.ca/~amg/expertise/remote_field.html> [cited November 21, 2002]

Anderson, R. 2003. The Gas Industry Perspective. Presented at the Workshop on Non-Interruptive Condition Assessment Inspection Devices for Water Transmission Mains, Haworth, N.J., March 12–14, 2003.

AOPL (Association of Oil Pipelines). 2000. Proposals for Mandatory Pipeline Integrity Tests [Online]. Available: <http://www.aopl.org/pubs/pdf/IntegrityTests.pdf> [cited October 17, 2002]

Atherton, D.L. 2000. Electromagnetic Method for Non-destructive Testing of Prestressed Concrete Pipes for Broken Prestressing Wires. U.S. Patent 6,127,823.

Austin, R. 1997. In-situ NDE of Buried Piping for Water Distribution [Online]. Available: <http://www.ntiac.com/news/news9705.html> [cited November 21, 2002]

Battelle Memorial Institute. 2002. Overview of Remote Field Eddy Current Techniques [Online].Available:<http://www.battelle.org/pipetechnology/DOTReport/links_direct/RFEC_Overview.htm> [cited November 21, 2002]

Bickerstaff, R.J., M. Vaughn, G. Stoker, M. Hassard, and M. Garrett. 2002. Review of Sensor Technologies for In-line Inspection of Natural Gas Pipelines [Online]. Available: <http://www.netl.doe.gov/scng/publications/t&d/tsa/71702.pdf> [cited November 21, 2002]

BJ Services Co., 2002. Available:http://www.bjservices.com/website/pps.nsf.

Bush, C., 2003. "Greater Cincinnati Waterworks Experience with Pipe Assessment," Presented at the Workshop on Non-Interruptive Condition Assessment Devices for Water Transmission Mains, Haworth, N.J., March 12-14, 2003.

Code of Federal Regulations 40 CFR Part 141.2 <http://www.access.gpo.gov/nara/cfr/cfrhtml 00/title 40/40cfr141 00.html

Conroy, P.J. 2003. The WRc Perspective – A Global Approach to Condition Assessment in Asset Management. Presented at the Workshop on Non-Interruptive Condition Assessment Devices for Water Transmission Mains, Haworth, N.J., March 12–14, 2003.

Crocket, P.A., and R. Maguire. 1999. Pipeline Failure Management. In *Ageing Pipelines; Optimizing the Management and Operation – Low Pressure – High Pressure*. Edited by IMechE: Professional Engineering Publishing.

Cromwell, III, J.E., H. Reynolds, N. Pearson, Jr., and M. Grant. 2002. *Costs of Infrastructure Failure*. Denver, Colo.: AwwaRF and AWWA.

CSIRO, 2004. Manufacturing and Infrastructure Technology, Available at <http://cmit.csiro.au/research/urbanwater/mouws/common2.cfm>

Deb, A.K., F.M. Grablutz, Y.J. Hasit, J.K. Snyder, G.V. Loganathan, and N. Agbenowsi. 2002a. *Prioritizing Water Main Replacement and Rehabilitation*. Denver, Colo.: AwwaRF and AWWA.

Deb, A.K., Y.J. Hasit, H.M. Schoser, and J.K. Snyder. 2002b. *Decision Support System for Distribution System Piping Renewal*. Denver, Colo.: AwwaRF and AWWA.

Deutsch, W.A.K., M. Platte, H. Schäfer, and P. Müller. 2002. Ultrasonic Plant Supervision in the Petrochemical Industry: A Wall-Thickness Measuring System for Fixed Installation On Endangered Piping Locations [Online]. Available:<http://www.ndt.net/article/ecndt02/28/28.htm> [cited October 14, 2002]

Dickson, D.T. 2000. *ECAT*. Cramlington, United Kingdom: AESL.

Dickson, D.T. 2002. *S19 Tool*. Cramlington, United Kingdom: AESL.

Dickson, D.T. 2003. General Parameters for Technology. Presented at the Workshop on Non-Interruptive Condition Assessment Inspection Devices for Water Transmission Mains, Haworth, N.J., March 12–14, 2003.

Eiswirth, M., and L.S. Burn. 2001. New Methods for Defect Diagnosis of Water Pipelines. Proceedings 4th International Conference On Water Pipeline Systems, York, U.K., 28-30 March 2001, pp. 137–150.

Fordyce, I.S., and M. Kidd. 1999. Pipeline Operation and Maintenance. In *Ageing Pipelines; Optimizing the Management and Operation – Low Pressure – High Pressure*. Edited by IMechE: Professional Engineering Publishing.

Geehman, C.J. 1999. Prioritizing Water Main Renewals [Online]. Available: <http://www.masterplumbers.com/plumbwatch/pipes99/watermain.doc> [cited October 11, 2002]

Glozzy, J. 2003. Pipeline Inspection Experience. Presented at the Workshop on Non-Interruptive Condition Assessment Inspection Devices for Water Transmission Mains, Haworth, N.J., March 12–14, 2003.

Hartman, W.F., K. Karlson and R. Brander. 2002. Waterline Restoration Based on Condition Assessment – A Case Study. Presented at the Distribution and Plant Operations Conference, Nashville, Tenn., September 7–10, 2002.

Heavens, J.W. The Trenchless Renovation of Potable Water Pipelines [Online]. Available: <http://www.insituform.com/resourceroom/Rr2_03.pdf> [cited January 29, 2003]

Hecht, A., R. Bauer, F. and Lindemeier. 1998. On-Line Radiographic Wallthickness-Measurement of Insulated Piping in the Chemical and Petrochemical Industry [Online]. Available: <http://www.ndt.net/article/ecndt98/101/101.htm> [cited October 11, 2002]

Hodges, D. 2001. Good forecast for pipelines. *Water and Waste Treatment*, 44(12):17.

iMAR GmBH. 2003. Inertial Measuring Technique [Online]. Available: <http://www.imar-navigation.de/download/prosp_2003_for_pdf.pdf> [cited January 31, 2003]

INGENIUM, and IPWEA (Association of Local Government New Zealand, Inc. and Institute of Public Works Engineering of Australia). 2002. *International Infrastructure Management Manual Version 2.0 (U.K. Edition)*. Thames, New Zealand: INGENIUM.

Insituform®. 2002. City of Indianapolis, Indiana, Invests $65 Million in Infrastructure with "Building Better Neighborhoods" Sewer Rehabilitation Project [Online]. Available : <http://www.insituform.com/resourceroom/resource_article13.html> [cited November 28, 2002]

Jackson, R.Z., C. Pitt, and R. Skabo. 1992. Recommendations. In *Nondestructive Testing of Water Mains for Physical Integrity*. Denver, Colo.: AWWARF and AWWA.

Jones, D.G. Extending the Performance of Ageing Pipelines [Online]. Available: <http://www.piigroup.com/reference_library/papers/pdfs/integrity/aging_pipelines.pdf> [cited November 21, 2002]

Kirmeyer, G.J., W. Richards, and C.D. Smith. 1994. *An Assessment of Water Distribution Systems and Associated Research Needs*. Denver, Colo.: AWWARF and AWWA.

Kopp, G. 2003. Global Condition Assessment Solutions. Presented at the Workshop on Non-Interruptive Condition Assessment Devices for Water Transmission Mains, Haworth, N.J., March 12–14, 2003.

Lebsack, S. 2002. Non-Invasive Inspection Method For Unpiggable Pipeline Sections. *Pipeline and Gas Journal*, June 2002:58

Legislative Council of Hong Kong. 1995. HEAD 194 – Water Supplies Department [Online]: Available: <http//www.legco.gov.hk/yr95-96/english/fc/fc/papers/fc031165.htm> [cited February 10, 2003]

Makar, J.M, and Y. Kleiner. 2000. *Maintaining Water Pipeline Integrity*. Ottawa, Ont.: National Research Council Canada.

May, A., and D.L. Atherton. 2002. Multi-frequency electromagnetic inspection of pre-stressed concrete cylinder pipe. *Insight*, 44(2):80–84.

Mergelas, B., and D.L. Atherton. 1999. Inspection of Prestressing Windings in Concrete Pressure Pipe. Presented at NACE International's 54th Annual Conference, San Diego, Calif. April 25–30, 1999.

Mergelas, B., and X. Kong. 2001. Executive Summary. In *Electromagnetic Inspection of Prestressed Concrete Pipe*. Denver, Colo.: AWWARF and AWWA.

Morgan, B., and D. Heath. 1999. The Application of Risk-Based Approaches to Common Pipeline Issues. In *Ageing Pipelines; Optimizing the Management and Operation – Low Pressure – High Pressure*. Edited by IMechE: Professional Engineering Publishing.

Morrow, M.J. 2003. Non Intrusive Condition Assessment of Water Pipes. Presented at the Workshop on Non-Interruptive Condition Assessment Inspection Devices for Water Transmission Mains, Haworth, N.J., March 12–14, 2003.

Neukrug, H. 2001. Statement on Drinking Water Needs and Infrastructure. Presented to the Environment and Hazardous Materials Subcommittee on Energy and Commerce, United States House of Representatives, Washington, D.C., March 28, 2001.

Oligo, O., L. Catalano, W. Worthington. 2003. *Validation of PCCP Pipeline Non-Destructive Test Results*. Mississauga, Ont.: The Pressure Pipe Inspection Company Ltd.

PII Pipeline Solutions. Are Pipelines Important? [Online]. Available: <http://www.piigroup.com/corporate_overview/pipelines/pipelines.html> [cited October 7, 2002]

Plotnikov, A., and L. Clapham. 2002. Stress Effects and Magnetic NDE Methods for Pipeline

PPIC, Pressure Pipe Inspection Company, Remote Field Eddy Current/Transformer Coupling, Available:http://www.ppic.on.ca/PDF/Marketing%20Material/rfec.pdf.

Rajani, B., J. Makar, S. McDonald, C. Zhan, S. Kuroaka, C. Jen, and M. Viens. 2000. Executive Summary. In *Investigation of Grey Cast Iron Water Mains to Develop a Methodology for Estimating Service Life*. Denver, Colo.: AWWARF and AWWA.

Randall-Smith, M., A. Russell, and R. Oliphant. 1992. *Guidance Manual for the Structural Condition Assessment of Trunk Mains*. Medmenham, Buckinghamshire, United Kingdom: WRc

Rasmussen, H.H., H. Kristensen, and L. Jeppesen. 1998. NDT and Heat Exchanger Tubes [Online]. Available: <http://www.ndt.net/article/ecndt98/chemical/117/117.htm> [Cited October 11, 2002]

Raymond, D. 2002. Cathodic Protection in Cold Climate Valve Case Studies Water Division [Online]. Available: <http://www.corrosionsource.com/corrosioneering/journal/Jul02_Raymond/Jul02_Raymond_1.htm> [cited December 2, 2002]

Reed, C., A. Robinson, and D. Smart. 2003. *Techniques for Monitoring Structural Behavior of Piping Systems*. Denver, Colo.: AWWARF and AWWA.

Robers, M.A., and R. Scottini. 2002. Pulsed Eddy Current in Corrosion Detection [Online]. Available: <http://www.ndt.net/article/ecndt02/251/251.htm> [cited October 14, 2002]

Roberts, R.D. 1999. Laser Profilometry as an Inspection Method for Reformer Catalyst Tubes [Online]. Available: <http://www.ndt.net/article/v04n02/roberts4/roberts4.htm> [cited October 11, 2002]

Roubal, M. 2003. *Pipeline Condition Assessment Techniques for Ferrous Pipelines*. Victoria, Australia: Rock Solid Group Pty. Ltd.

Sægrov, S., W. Schilling, J. Røstrum, L. Tuhovcak, P. Eisenbeis, R. Herz, P. LeGauffre, J.M. Baptista, P. Conroy, V. deFederico, S. Mazzacane, and M. Schiatti. 2002. Computer Aided Rehabilitation of Water Networks (CARE-W). Presented at the NODIG 2002 Conference, Copenhagen, Denmark, May 29, 2002.

Sutherland, A. 2000. Fact Sheet 10 – Guaranteed Minimum Standards: Water Industry Commissioner for Scotland [Online]. Available: <http://www.watercommissioner.co.uk/FS%2010%20GMS.html> [cited October 22, 2002]

Sutton and East Surrey Water Plc. 2000. Appendix 6.1: The quality enhancement/capital maintenance overlap. In *A report on the references under sections 12 and 14 of the Water Industry Act 1991*. London, United Kingdom: Competition Commission.

Tawfiq, K. 2002. Lecture #1 Soil and Soil Engineering [Online]. Available: <http://www.eng.fsu.edu/~tawfiq/soilmech/lecture.html> [cited November 27, 2002]

Trench, S.J., and C.A. Sturbitts. 1999. *The Environmental and Safety Programs of Oil Companies: A Prevention Ethic*. Washington, D.C.: Association of Oil Pipe Lines.

USDOE (United States Department of Energy). 2003. EH-41 Environmental Law Summary: Safe Drinking Water Act [Online]. Available: <http://tis.eh.doe.gov/oepa/law_sum/sdwa.cfm> [cited April 3, 2003]

USEPA (United States Environmental Protection Agency). 1999. *25 Years of the Safe Drinking Water Act: History and Trends*. Washington, D.C.: United States Environmental Protection Agency.

USEPA (United States Environmental Protection Agency). 2003a. About Superfund [Online]. Available: <http://www.epa.gov/superfund/about.htm> [cited April 3, 2003]

USEPA (United States Environmental Protection Agency). 2003b. Superfund Frequently Asked Questions [Online]. Available: <http://cfpub.epa.gov/superapps/index.cfm/fuseaction/faqs.viewAnswer/question_id/138/category_id/12/faqanswr.cfm> [cited April 3, 2003]

USEPA (United States Environmental Protection Agency). 2003c. Part 141 National Primary Drinking Water Regulations [Online]. Available: <http://www.access.gpo.gov/nara/cfr/cfrhtml_00/Title_40/40cfr141_00.html> [cited August 14, 2003]

USEPA (United States Environmental Protection Agency) "Drinking Water Needs Survey" <http://www.epa.gov/safewater/needs.html

USEPA (United States Environmental Protection Agency) "The Clean Water and Drinking Water Infrastructure Gap Analysis <http://www.epa.gov/owm/gapfact.pdf
Vollman, K.W. 1998. The Emerging Context for the Physical Regulation of Pipelines. Presented at International Pipeline Conference, Calgary, Alta., June 9, 1998.
Wall, M. and J. Lilley. 2002. The Application of POD Data in Inspection Planning using the Inspection Value Method. Presented at 3rd European-American Workshop on Reliability of NDE and Demining, Berlin, Germany, September 12, 2002.
Water Industry Act. 1991. Sections 54, 57, 58, 65 and 70. United Kingdom.
Wayman, M. 2003. Life Prediction and Condition Assessment for Large Diameter Water Pipelines. Cramlington, United Kingdom: AESL.
Winn, K. 2003. Acoustic Emission Monitoring of Prestressed Concrete Cylinder Pipe (PCCP). Presented at the Workshop on Non-Interruptive Condition Assessment Inspection Devices for Water Transmission Mains, Haworth, N.J., March 12–14, 2003.
WRc. 2003. Sahara Leak Location System for Trunk Mains [Online]. Available: <http://www.wrcplc.co.uk/downloads/Sahara.pdf> [cited July 23, 2003]

ABBREVIATIONS

”	inch
£	Pounds Sterling
$	Unitcd Statcs Dollars
<=	less than or equal to
A–C	asbestos cement
ACFM	alternating current field measurement
AE	acoustic emission
AESL	Advanced Engineering Solutions Limited
AG	Aktiengesellschaft (company, vs. corporation)
AL	acoustic leak detection
AMG	Applied Magnetics Group
ARARs	Applicable or Relevant and Appropriate Requirements
AWWA	American Water Works Association
AwwaRF	Awwa Research Foundation
BE	broadband electromagnetic
BEM	broadband electromagnetic
°C	degrees Celsius
CARE-W	Computer Aided Rehabilitation of Water Networks
CCTV	closed circuit television
CDSS	comprehensive decision support system
CEO	Chief Executive Officer
CF	consequences of failure
CFR	Code of Federal Regulations
CI	cast iron
CLCI	cement lined cast iron
cm	centimeter
d_e	equivalent diameter
DEFRA	Department of Environment, Food and Rural Affairs
DI	ductile iron
DOE	Department of Energy
DTG	dynamically tuned gyros
DWI	Drinking Water Inspectorate
E	external
EA	Environment Agency (U.K.)
EC	eddy current
ECAT	external condition assessment tool
ECP	embedded cylinder pipe
°F	degrees Fahrenheit

FOG	fiber-optic gyros
ft	foot
GASB	Governmental Accounting Standards Board
GIS	geographic information system
GmBH	Gesellschaft mit Beschränkter Haftung (Limited Liability Company)
GPR	ground penetrating radar
GPS	Global Positioning System
HDPE	high-density polyethylene
HK$	Hong Kong Dollars
I	internal
ID	inside diameter
IE	impact echo
ILI	in-line inspection
IMU	Inertial Measurement Unit
Inc.	Incorporated
INS	inertial navigation system
IVM	Inspection Value Method
km	kilometer
LCP	lined cylinder pipe
LOS	levels of service
LRUT	long-range ultrasonic testing
Ltd.	Private Limited Company
LU	long-range ultrasonic testing
LUC	land use category
m	meter
MCLGs	Maximum Contaminant Limit Goals
MCLs	Maximum Contaminant Limits
MDPE	medium-density polyethylene
MEMS	micromachined sensors
MF	magnetic flux leakage
MFL	magnetic flux leakage
MHz	megahertz
min	minute
mm	millimeter
N	no
N/A	does not apply
NDE	non-destructive evaluation
NDT	non-destructive testing
NPL	National Priority List

NPV	Net Present Value formulation
NRC	National Research Council Canada
NTNCWS	non-transient non-community water system
NWTC	Northumbrian Water Technical Centre Limited
OD	outside diameter
ODRC	optimized depreciated replacement cost
OFWAT	Office of Water Services
ORC	optimized replacement cost
ΔP	reduction of the probability of failure
PAC	Project Advisory Committee
PCCP	prestressed concrete cylinder pipe
plc.	Public Limited Company
PPIC	Pressure Pipe Inspection Company Ltd.
PSC	prestressed concrete
Pty.	Party
PVC	polyvinyl chloride
PVG	piezo-vibrating gyros
RF	remote field eddy current
RFEC	remote field eddy current
RFEC/TC	remote field eddy current – transformer coupling
RLG	ring laser gyros
RMPS	rapid magnetic permeability scanning
RSG	Rock Solid Group Pty. Ltd.
RUL	remaining useful life
S	steel
SCC	stress corrosion cracking
SDWA	Safe Drinking Water Act 1974
SEWL	South East Water
SF	safety factor
SP	seismic pulse echo
TAGs	Technical Assistance Grants
TOFD	time of flight diffraction
TR	remote field eddy current – transformer coupling
TV	closed circuit television
TWS	transient non-community water system
U.K.	United Kingdom
uPVC	non-plasticized polyvinyl chloride
U.S.	United States
U.S.A.	United States of America
USDOE	United States Department of Energy

USEPA	United States Environmental Protection Agency
UT	ultrasonic testing
VOCs	volatile organic compounds
WATERFOWL	WRc Analysis of Techniques for Effective Rehabilitation Financed Over Whole Life
WRAP	Watermain Renewal Analyser & Prioritisor
WRc	Water Research Centre
WT	wall thickness
Y	yes

IWA Publishing's authorised EU representative for General Product Safety Regulations is Diane D'Arras, 15 rue Duret, 75116 Paris, France, e-mail: safety@iwap.co.uk.

Printed and bound by CPI Group (UK) Ltd, Croydon, CR0 4YY
17/09/2026
02218804-0018